# New Directions in Behavioral Biometrics

# New Directions in Behavioral Biometrics

## Khalid Saeed

with Marcin Adamski, Tapalina Bhattasali,
Mohammad K. Nammous, Piotr Panasiuk,
Mariusz Rybnik, and Soharab H. Shaikh

CRC Press
Taylor & Francis Group
Boca Raton   London   New York

CRC Press is an imprint of the
Taylor & Francis Group, an **informa** business

CRC Press
Taylor & Francis Group
6000 Broken Sound Parkway NW, Suite 300
Boca Raton, FL 33487-2742

Printed on acid-free paper
Version Date: 20160531

International Standard Book Number-13: 978-1-4987-8462-7 (Hardback)

### Library of Congress Cataloging-in-Publication Data

Names: Saeed, Khalid (Computer scientist), author.
Title: New directions in behavioral biometrics / Khalid Saeed with Marcin
Adamski, Tapalina Bhattasali, Mohammad K. Nammous, Piotr Panasiuk, Mariusz
Rybnik, and Soharab H. Shaikh.
Description: Boca Raton : CRC Press, Taylor & Francis Group, 2017. | Includes
bibliographical references and index.
Identifiers: LCCN 2016016651 | ISBN 9781498784627
Subjects: LCSH: Biometric identification. | Behaviorism (Psychology)
Classification: LCC TK7882.B56 S236 2017 | DDC 006.2/48--dc23
LC record available at https://lccn.loc.gov/2016016651

**Visit the Taylor & Francis Web site at**
**http://www.taylorandfrancis.com**

**and the CRC Press Web site at**
**http://www.crcpress.com**

Printed and bound in the United States of America by Publishers Graphics,
LLC on sustainably sourced paper.

To my assistants and coauthors who have always
proved smart, outstanding, and prominent.

# Contents

# Foreword

I met Professor Saeed, the principal author of this book, for the first time in 2009 at AGH University of Science and Technology, Kraków, Poland. I was invited as one of the keynote speakers and a researcher of behavioral biometrics in the International Multi-Conference on Biometrics and Kansei Engineering. My first impression of him was a true professor in an elegant suit, with high-quality speech, passionate approach to research, and unfailing kindness. Although 7 years have already passed, his humanity is abiding, and I have ever since constantly communicated with him as a researcher, a university educator, a journal editor, and an international conference chair.

Professor Saeed has achieved a great accomplishment by writing this book: *New Directions in Behavioral Biometrics*. Publishing this book has been a delightful event for me and many biometrics researchers. Based on the great achievement of his research team, this book includes a readable introduction, beneficial literature reviews, specific approaches, detailed algorithms, and useful experimental results of behavioral biometrics such as signature recognition, keystroke dynamics, gait analysis, and voice recognition.

Biometric technology using physiological characteristics has been widely applied, and many products that employ this technology are very popular, whereas biometric technology using behavioral characteristics has not received considerable attention. However, from a different point of view, research of the behavioral biometrics possesses the possibility of success and advancement. For a university student, a researcher, or a developer who is researching biometrics or developing an application system, this book will provide a clear path of basic information on behavioral biometrics and many hints for new discovery, all coming out of the extensive research experience of the author.

If, in the near future, a technology of identifying an individual based on writing style is developed, Professor Saeed will be easily identified because this book includes his passionate approach and a high degree of completion.

**Nobuyuki Nishiuchi**
*Tokyo Metropolitan University*

# Preface

This book is the result of research conducted with major members of my international biometrics team on some selected topics in behavioral biometrics: a field of study related to the way people behave. As the reader might recall, the science of biometrics deals with biological measurement—describing and measuring human features for the sake of identity recognition (personal authentication and identification or verification, or both). Biometrics is known in two categories: physiological and behavioral. *Physiological biometrics* is beyond the scope of this book, but some selected examples have been provided in another book by the author (see *Biometrics and Kansei Engineering*, Springer, New York, 2012). This book, though, will focus on the most popular and emblematic examples of behavioral biometrics, namely, signature (the way we write), dynamics of keystrokes or touchscreen use (how we use the keyboard or touchscreen to enter text), and gait (how we move our legs to walk). Chapter 5 will deal with speech and speaker recognition, that is, the manner and art of combining vocal tracts, tongue, and lips to utter sounds and say words, which is sometimes considered to be a combination of physiological and behavioral features. All these human features are unique; they differ from one person to another, and that is why they can be used to distinguish people.

The content of the book is divided among five chapters. Chapter 1 is a general introduction to the subject, showing the nature of each behavioral feature, and discussing how they are collected and prepared for measuring. It also discusses the advantages and disadvantages of using such features for user authentication or personal verification.

These chapters reflect the authors' extensive research on the different examples of biometrics. We have published many detailed works and consequently received valuable feedback from the reviewers, readers, and conference participants whenever a research work was presented for discussion. Hence, I hope this book appears in a well-organized and readable form to students and researchers.

As biometrics science is developing fast, I can say there is no single written book that could comprise all known behavioral biometrics aspects and that would incorporate all existing topics of this new scientific discipline. Apart from the small number of standard academic books on behavioral biometrics, students usually make use of available research papers and some special book briefs on some biometrics topics and aspects. However, researchers still seem to face difficulties introducing new, essential methods in behavior recognition with good enough results. This mainly comes from the basic characteristics of behavioral biometrics features, especially their *low repeatability*. This causes low accuracy and hence less system efficiency and a lower recognition rate when behavioral biometrics is considered alone. Still, researchers will always hope their methods will find their right place when technological advances produce sensors that are able to capture unique features that can give a sufficient unique description for different objects. Until then, the best way a behavioral biometrics feature can perform in a recognition system is by being a part of a fusion or hybrid system. In such systems, biometric features can play a partial role with other physiological features, as fingerprints provide another method of user authentication along with the PIN code, for example.

Many successful examples, such as signature, keystroke dynamics, gait, and voice feature recognition, are being developed with higher recognition rates, and this gave rise to the need for a book that discusses the most important and practical behavioral biometrics features. The methods and algorithms given on these features in this book have been checked for the highest possible recognition rates.

In Chapter 1, the reader can find introductory information about biometrics and the known behavioral features of a human.

Chapter 2, which is one of the most comprehensive chapters, introduces the basic achievements in human recognition by use of signatures—with only one reference signature, a user can be verified/personified. Dr. Marcin Adamski had proved this important result in his PhD thesis under my supervision in 2010. Since then, he has been working on algorithm improvement to reach the most accepted practical approach in everyday applications.

Chapter 3 is about keystroke dynamics—personalization based on how a user types on a keyboard. This *human–computer interaction* seems to be effective in verifying our contacts without the necessity or the possibility of seeing them. In addition to its practical use in patient–physician interactions, this method allows us to identify the age interval of the user, which is important for parents who wish to know who their children are having contact with. In their PhD works, Tapalina Bhattasali and Piotr Panasiuk have worked out simple patient–physician remote contact models, in which user passwords are accompanied by keyboard strike dynamics.

Chapter 4 deals with gait recognition for personal verification. It introduces the most recent research results of Dr. Soharab H. Shaikh, who worked on gait analysis as one of the basic parts of his PhD work.

In Chapter 5, voice recognition for personal verification and identification has been discussed on the basis of the heuristic studies of Mohammad K. Nammous as part of his PhD study.

This book contains practical examples, illustrations, and simple algorithms or algorithmic descriptions for students to easily implement in their computers. Moreover, the contents have been structured in such a way that each chapter can be read separately without the need to go back to any other chapter. They are self-contained and each chapter covers a separate subject area.

All my coauthors have been or still are my students, assistants or coworkers, and researchers, a fact that I have been proud of.

**Khalid Saeed**
*Bialystok University of Technology*

# Acknowledgments

I thank the book reviewers for their invaluable comments to make the book a success.

I am indebted to Rich O'Hanley of Taylor & Francis Group for his support and steady encouragement to write and publish this book. I also am thankful to his publishing team, in particular, Richard Tressider (project editor) and Balagowri Murugan (project manager of Lumina Datamatics) for their great efforts and insightful comments, which made this book appear in a clear and extremely readable form.

# 1

# INTRODUCTION
# TO BEHAVIORAL BIOMETRICS

Biometrics refers to the study of biological characteristics. It comes from Greek words "bios" (implies life) and "metricos" (implies measuring/to measure). Biometrics can be considered as use of physiological or behavioral characteristics in an automated way to determine identity. The identity verification is performed through the measurement of physiological or behavioral characteristics of an individual. Researchers have proposed a number of biometric techniques for human identification and authentication based on fingerprint, palm print, hand geometry, face, ear, iris, retina, voice, signature, body odor, and so on.

Biometric traits are almost statistical in nature. The system is likely to be unique and reliable as much as data are available from sample. It can work on various modalities based on the measurements of individual's body and features, and behavioral patterns. The modalities are classified according to individual's biological traits. The biometric modalities mainly fall under following types:

- Physiological
- Behavioral
- Hybrid

Physiological category includes the features we are born with. This modality is based on the shape and size of the body. Examples are [1–3] as follows:

- Fingerprint recognition
- Hand geometry recognition
- Facial recognition
- Iris recognition

- Retinal scanning
- DNA recognition

Behavioral category deals with the features we learn in our life as a result of our interaction with the environment and the nature. This modality is related to change in human behavior over time. Examples of this category are

- The way we walk (gait).
- The way we write (signature).
- The way we speak or say a word (voice).
- The way we type on a machine (keystroke dynamics).
- Many other ways of our response to the natural events around us and the way we react to or respond.

Hybrid modality includes both traits, where the traits are depending upon physical as well as behavioral changes. As for example, voice recognition may be considered as hybrid modality as it depends on size and shape of vocal cord, nasal cavities, mouth cavity, shape of lips, and so on, and the emotional status, age, illness (behavior) of a person.

Hybrid modality is also considered as a type of multimodality (more than one mode involved). In this book, the concept of behavioral biometrics is presented briefly on the basis of some selected features like signature, keystroke dynamics, gait, and voice. A brief overview of behavioral biometrics is presented in this chapter.

## 1.1 Behaviometrics

The word "behaviometrics" derives from the terms "behavioral" and "biometrics" [4]. Behavioral refers to the way an individual behaves. Behaviometrics, or behavioral biometrics, is a measurable behavior used to recognize or verify the identity of a person. Behaviometrics focuses on behavioral patterns rather than physical attributes.

### 1.1.1 How It Works

Each person has a unique pattern—how they interact with computing devices by using keyboard, mouse, and graphical user interface (GUI). The study of the user's unique nature in this regard is known as behaviometrics.

A human behavioral pattern consists of a variety of different unique behaviors—all are mixed together into a larger unique profile. Since unique behaviometric pattern of every person is formed not only by biometric features, but is also influenced by social and psychological means, it is just impossible to copy somebody else's behavior.

The behavioral pattern of the person is compared with the stored pattern. Matching scores of similarities for those users are recognized and the software calculates the possibility of accurate identification of users.

The key features of behavioral biometrics are given below.

- Security of applications like user authentication and intrusion detection may be enhanced by behavioral biometrics with very low impact on the users.
- Behavioral biometrics is highly sensitive to the means of implementation, for example, keystroke dynamics depend on the type of used keyboard.
- Behavioral biometrics is most useful in multimodal systems (where more than one type of biometrics is used at the same time) compared to unimodal systems (where only one type of biometrics is used at a time).
- It may be vulnerable to several spoofing attacks [5].

A comprehensive review of different biometric technologies including theory and applications can be found in [1]. A survey of different techniques on behavioral biometrics is summarized in [6]. Behavioral biometrics is popularly used in information security context to identify individuals by using unique features of activities they perform either consciously or unconsciously. In recent times, it has been observed that behavioral biometric data are used for a number of interesting applications. Researchers have proposed methodologies for speaker recognition by tracking movements of lips [7], biometric verification using motions of fingers [8], and extracting biometric features of voice [9] for person identification. A promising application of biometrics is artimetrics, where biometric traits are used for authenticating artificial entities like industrial robots, intelligent software agents, and virtual-world avatars [9]. Biometric data are also used for enhancing the security of cryptographic systems. New algorithms are developed by researchers for filtering biometrics. Performance evaluation of systems is very important for the following reasons.

- Quality of system must be precisely quantified to be used in real context. To determine whether it fulfills the requirements of a specific application based on logical or physical access, context of use, efficiency, and robustness of the logic must be defined.
- Comparison of different biometric modalities is essential to analyze their relative merits and demerits.
- Performance evaluation is also necessary in order to facilitate research in this field.

Evaluation techniques are used to quantify the performance of behavioral biometric systems. A reliable evaluation method is needed in order to analyze advantage of the system.

### 1.1.2 Major Benefits

Behaviometrics can provide information security solutions by using the nature of individual. It is extremely hard to replicate, which makes it the ultimate solution against identity theft. It is not possible that any unauthorized user could access a computer with confidential information, either by hacking the password or logging in with stolen credentials or accessing a logged on computer. As a result, intrusion can be prevented.

As for example, it is possible to recognize and confirm the identity of a person by analyzing how the user works with the keyboard (typing rhythm), mouse movements (acceleration time, click frequencies), and graphical interface interaction (using programs).

- While many popular security solutions require the user to perform additional tasks, behaviometrics does not interfere with the normal work flow. Simple use of computing device in the everyday work makes the software increasingly more efficient and the confidential information more secure.
- Behaviometrics will allow workstations to be secure even after the user has logged on to the system. Even if the user leaves

the workstation and forgets to sign out, computing device stays protected.

- Existing token-based products (such as passwords and smart cards) can be duplicated or stolen, whereas user's behavior is unique and very difficult to copy.

## 1.2 What Is Special about Behavioral Biometrics Data Acquisition?

The special aspects of behavioral biometrics data acquisition are presented below.

- It offers increased convenience in data acquisition, because there is no requirement for dedicated or special hardware. As a result, it is also considered as cost-effective.
- Most of the data are acquired through machine-based interactions.
- These traits need to be easily verifiable and identifiable.
- Input data depend on the permanence and distinctiveness metrics.
- It does not introduce delays in operations and are implemented silently. It is mostly used in online platforms. Their acceptance level in the society is high.

## 1.3 Behavioral Biometrics Features

A biometric system may include different phases. Two phases are mainly considered during the use of a biometric system. A working model of data acquisition is defined in enrollment phase of an individual. Verification phase uses this model to make a decision about an individual. The performance evaluation of a biometric system generally considers the quality of the input data and the output result.

In accomplishing their everyday tasks [1], people employ different strategies, use different styles and apply unique skills and knowledge. One of the defining characteristics of a behavioral biometric is the incorporation of time dimension as a part of the behavioral signature. The measured behavior has a beginning, duration, and an end. Researchers attempt to quantify behavioral traits exhibited by users and use resulting features to verify identity efficiently.

Behavioral biometrics provides a number of advantages. They can be collected without the knowledge of the user. It becomes very cost-effective as collection of behavioral data often does not require any special hardware. Data acquisition devices include computer, keyboard, mouse, stylus, touch screen, microphone, camera, credit card, and scanner to capture most frequently used behavioral biometrics. Although most of the behavioral biometric traits may not be unique enough to provide reliable human identification, it is observed that they can provide sufficiently high accuracy in identity verification.

Behavioral biometric systems are requirement specific. Many characteristics make them difficult to analyze their performance [10]:

- Biometric template generally contains temporal information.
- This type of template can change with time. It means that the biometric template can be quite different compared to the one obtained after the enrollment phase.
- The behavior of biometric characteristic can be very different for an individual given its age, culture, and experience.

The evaluation of system is often analyzed by considering a variety of users.

Benchmark definition has high impact on the performance evaluation of biometric systems. A benchmark database may include either real biometric templates or synthetic ones. The definition of synthetic templates is easier for behavioral biometric data such as keystroke dynamics, voice, lip movements, mouse dynamics, and signature dynamics. For behavioral biometric modalities such as keystroke dynamics, voice, or gait, the associated template can vary for individuals at different ages. As a consequence, the benchmark database must include all the variability of biometric templates to represent real applications.

Behavioral biometrics and related technologies have potential to improve diverse areas such as mobile commerce, real application analysis, risk, and financial analysis. It is used for user modeling that aims at creating a representation of the user for the purpose of customizing service suitable for the user.

## 1.4 Classification of Behavioral Biometrics Traits

Behavioral biometrics can be used in an information security context to identify individuals by using unique features of activities they perform either consciously or unconsciously. Behavioral biometrics can be classified into several categories based on the type of information about the user being collected (Figure 1.1).

Behavioral biometrics traits are entirely dependent on behavioral nature of human beings. It measures human behavior which is not directly focusing on measurements of body parts.

Motor skill of a human being has an ability to utilize muscles. Muscle movements rely upon the proper functioning of the brain, skeleton, joints, and nervous system. Therefore, motor skills indirectly reflect the quality of functioning of such systems, making person verification possible. Most of the motor skills are learned, not inherited. Definition for motor skill is adopted based on behavioral biometrics, for example, "kinetics," which are based on unique and stable muscle actions of the user while performing a particular task.

Authorship-based biometrics is based on examining a piece of text or a drawing produced by a person. Verification is accomplished by observing style peculiarities typical to the author of the work being examined, such as the used vocabulary, punctuation, or brush strokes.

Human–computer interaction (HCI)-based biometrics can be further subdivided into additional categories—indirect HCI-based interaction and direct HCI-based interaction. Indirect HCI-based biometrics includes the events that can be obtained by monitoring user's HCI behavior indirectly via observable low-level actions of computer software. Direct HCI-based interaction is again divided into

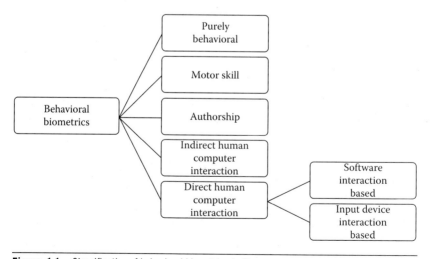

**Figure 1.1**   Classification of behavioral biometrics traits.

two categories. The first category consists of human interaction with input devices such as keyboards and mouse. The second category consists of HCI-based behavioral biometrics, which measures advanced human behavior such as strategy, knowledge, or skill exhibited by the user during interaction with different software (Figure 1.2).

> *Signature verification*: Here, users need to present handwritten text for authentication.
>
> *Keystroke dynamics*: It is a behavioral biometrics that relies on the way by which users can interact with keyboard. As a person interacts with keyboard, features are extracted and used to identify the user.
>
> *Gait analysis*: Gait is considered as biometrics controlled by muscle. Gait analysis is the systematic study of human motion, using the eye and the brain of observers, for measuring body movements, body mechanics, and the activity of the muscles.
>
> *Voice recognition*: Voice is used for either speaker identification or speaker authentication.

Other interesting examples include

- *Blinking pattern*: Time between blinks, how long the eye is held closed at each blink, physical characteristics of the eye while blinking
- *ECG*: Features of electromagnetic signals generated by the heart
- *EEG*: Features of electromagnetic signals generated by the brain

A number of techniques are limited to specific use-cases—for example, car driving style (to identify drivers), handgrip pressure patterns (for authentication to handheld devices including weapons), and credit

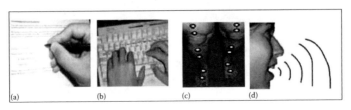

**Figure 1.2**    Behavioral traits: (a) signature, (b) keystroke, (c) gait, and (d) voice.

card usage patterns (credit card fraud detection). Important factors in the successful implementation of behavioral biometrics include

- *Equipment required*: This can vary from none at all (e.g., in the case of simple keystroke dynamics) to multiple cameras, EEG sensors, and so on.
- *Enrollment time*: The time required to train the system to recognize individual.
- *Persistence*: The time before an identifying feature changes in an individual after an initial training period of the system.
- *Obtrusiveness*: How much the system alters the normal experience of the identification subject.
- *Error rates*: As with all biometrics, error rates are analyzed according to
  - *False rejection rate (FRR)*: The percentage of individuals wrongly denied access to a system.
  - *False acceptance rate (FAR)*: The percentage of individuals wrongly authorized by a system.
  - *Equal error rate (EER)*: A measure often used to evaluate the accuracy of a biometric technology. It is the value of FRR and FAR when a system is tuned to have an equal FAR and FRR.

## 1.5  Properties of Few Behavioral Biometrics

### 1.5.1  Signature

In signature recognition, more emphasis is given on the behavioral patterns in which the signature is signed rather than its visibility in terms of graphics.

The behavioral patterns include the changes in the timing of writing, pauses, pressure, direction of strokes, and speed during the course of signing. It could be easy to duplicate the graphical appearance of the signature, but it is not easy to copy the signature with the same behavior the person shows while signing (Figure 1.3).

This technology consists of a pen and a specialized writing tablet, both connected to a computer for template comparison and verification. A high quality tablet can capture the behavioral traits such as speed, pressure, and timing while signing.

**Figure 1.3**   Signature recognition.

During enrollment phase, the candidate must sign on the writing tablet multiple times for data acquisition. The signature recognition algorithms then extract the unique features such as timing, pressure, speed, direction of strokes, important points on the path of signature, and the size of signature. The algorithm assigns different values of weights to those points. At the time of identification, the candidate enters the live sample of the signature, which is compared with the signatures in the database.

Signature verification is a widely accepted methodology for confirming identity. Two distinct approaches to signature verification are recognized based on the data collection approach. They are online and offline signature verification, also known as static and dynamic approaches. In the offline signature verification, the image of the signature is obtained using a scanning device. The offline approach utilizes the static features of the signature. With online signature verification, special hardware is used to capture dynamics of the signature. During online signature verification, signature characteristics are extracted as the user writes, and these features are used to immediately authenticate the user. Typically, specialized hardware is required, such as a pressure-sensitive pen or a special writing tablet. These hardware elements are designed to capture pen pressure, pen angle, and related information. During remote authentication, online approach is most suitable. Signature-related features can be classified into two groups—global and local. Global features include signing speed, signature bounding box, Fourier descriptors of the signature's trajectory, number of strokes, and signing flow [11]. Local features describe specific sample point in the signature and relationship between such points, for example, distance and curvature change between two successive points may be analyzed as well as $x$ and $y$ offsets relative to the first point on the signature trajectory, and critical points of the signature trajectory [11].

### 1.5.1.1  Constraints of Signature Recognition
- To acquire adequate amount of data, the signature should be small enough to fit on tablet and big enough to be able to deal with.
- The quality of the writing tablet decides the robustness of signature recognition enrollment template.
- The candidate must perform the verification processes in the same type of environment and conditions as they are at the time of enrollment. If there is a change, then the enrollment template and live sample template may differ from each other.

### 1.5.1.2  Merits of Signature Recognition
- Signature recognition process has a high resistance to impostors as it is very difficult to imitate the behavior patterns associated with the signature.
- It works very well in high amount business transactions.
- It is a noninvasive tool.
- There are no privacy rights issues involved.
- Even if the system is hacked and the template is stolen, it is easy to restore the template.

### 1.5.1.3  Demerits of Signature Recognition
- The live sample template is prone to change with respect to the changes in behavior while signing.
- User needs to get accustomed to using signing tablet. Error rate is high till it happens.

### 1.5.1.4  Applications of Signature Recognition
- It is used in document verification and authorization.

The concept of biometrics sketch authentication [11] is similar to the concept of signature recognition. This method is based on sketch recognition and a user's personal knowledge about the drawings content.

The system directs a user to create a simple sketch. Sketches of different users are sufficiently unique to provide accurate authentication. The approach measures user's knowledge about the sketch, which is only available to the previously authenticated user. Features like location and relative position of different primitives are taken as the profile of the sketch.

### 1.5.2  Keystroke Dynamics

During World War II, a technique known as Fist of the Sender was used by military to determine if the Morse code was sent by an enemy or ally based on the rhythm of typing. These days, keystroke dynamics is the easiest biometric solution to implement in terms of hardware [12].

This biometric analyzes candidate's typing pattern, rhythm, and speed of typing on a keyboard. The dwell time and flight time measurements are used in keystroke recognition.

*Dwell time*: It is the duration of time for which a key is pressed.
*Flight time*: It is the time elapsed between releasing a key and pressing the following key.

The candidates differ in the way they type on the keyboard as the time they take to find the right key, the flight time, and the dwelling time. Their speed and rhythm of typing also varies according to their level of comfort with the keyboard. Keystroke recognition system monitors the keyboard inputs thousands of times per second in a single attempt to identify users based on their habits of typing.

Typing patterns [11] are characteristic of each person. For verification, a small typing sample such as the input of user's password is sufficient, but for recognition, a large amount of keystroke data are needed and identification is based on comparisons with the profiles of all other existing users already in the system. Keystroke features are based on time durations between the keystrokes, flight times and dwell times, overall typing speed, frequency of errors (use of backspace), use of numpad, order in which user presses shift key to get capital letters, and possibly the force with which keys are hit for specially equipped keyboards. Keystroke dynamics is probably the most researched type of HCI-based biometrics [11].

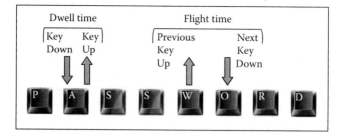

**Figure 1.4**  Keystroke verification.

There are two types of keystroke recognition.

- *Static*: It is one-time recognition at the start of interaction.
- *Continuous*: It is throughout the course of interaction (Figure 1.4).

### 1.5.2.1  Merits of Keystroke Recognition
- It needs no special hardware to track this biometric.
- It is a quick and secure way of identification.
- A person typing does not have to worry about being watched.
- Users need no training for enrollment or entering their live samples.

### 1.5.2.2  Demerits of Keystroke Recognition
- The candidate's typing rhythm can change because of tiredness, sickness, influence of medicines or alcohol, change of keyboard, and so on.
- There are no known features dedicated solely to carry out discriminating information.

Based on the idea of monitoring user's keyboard and mouse activity [11], system is developed for collecting GUI interaction-based data. Collected data allows generation of behavioral profiles of end-users. Such type of data may provide additional information not available from typically analyzed command-line data. Ideally, the collected data would include high-level detailed information about the GUI-related actions of the user such as left click on the start menu and so on. All collected data are time stamped and preprocessed to reduce the amount of data actually used for intrusion detection purposes.

### 1.5.2.3 Application of Keystroke Recognition

- Keystroke recognition is used for identification/verification. It is used with user ID/password as a form of multifactor authentication
- It is used for surveillance. Some software solutions track keystroke behavior for each user account without end-user's knowledge. This tracking is used to analyze if the account was being shared or used by anyone else than the genuine account owner. It is used to verify if some software license is being shared
- Application area:
  - Student identification for online examinations
  - User/employee identification for remote workstations
  - User authentication for network access
  - User authentication in e-commerce, online banking, and e-government

### 1.5.3 Gait

Gait is the manner of a person's walking. People show different traits while walking such as body posture, distance between two feet while walking, swaying, and so on, which helps to recognize them uniquely.

A gait recognition based on the analysis of video images containing candidate's walk. The sample of candidate's walk cycle is recorded by video. The sample is then analyzed for position of joints such as knees and ankles, and the angles made between them while walking.

A respective mathematical model is created for every candidate and stored in the database as a template. At the time of verification, this model is compared with the live sample of the candidate walk to determine its identity (Figure 1.5).

Gait [11] is a complex spatiotemporal motor-control behavior which allows biometric recognition of individuals at a distance usually from captured video. Gait analysis is used to assess, plan, and treat individuals with conditions affecting their ability to walk. It is also commonly used to identify posture-related or movement-related problems in people with injuries. It can distinguish one individual

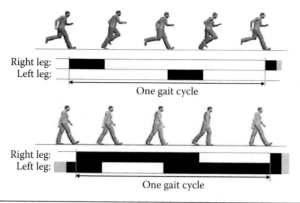

**Figure 1.5**  Gait recognition.

from others from various aspects about the human being (health, age, size, weight, speed, etc.) from his/her gait pattern. Typical features include amount of arm swing, rhythm of the walker, bounce, length of steps, vertical distance between head and foot, distance between head and pelvis, and maximum distance between the left and right foot [11]. Gait variation is based on changes in person's body weight, injuries, and so on.

### 1.5.3.1 Merits of Gait Recognition
- It is noninvasive.
- It does not need the candidate's cooperation as it can be used from a distance.
- It can be used for determining medical disorders by spotting changes in walking pattern of a person in case of Parkinson's disease.

### 1.5.3.2 Demerits of Gait Recognition
- For this biometric technique, developing model with complete accuracy is very difficult.
- It may not be as reliable as other established biometric techniques.

### 1.5.3.3 Application of Gait Recognition    It is well-suited for identifying criminals in the crime scenario.

*1.5.4 Voice*

Voice Recognition is also called speaker recognition. At the time of enrollment, the user needs to speak a word or phrase into a microphone. This is necessary to acquire speech sample of a candidate (Figure 1.6).

Speaker identification is one of the best researched biometric technologies. Verification is based on information about the speaker's anatomical structure conveyed in amplitude spectrum, with the location and size of spectral peaks related to the vocal tract shape. Speaker identification systems can be classified based on the freedom of what is spoken.

- *Fixed text*: The speaker says a particular word selected at enrollment.
- *Text dependent*: The speaker is prompted by the system to say a particular phrase.
- *Text independent*: The speaker is free to say anything he wants. Verification accuracy typically improves with larger amount of spoken text.

One of the principal tasks of the enrollment process is to register the person as a potential user of the biometric system. In a speaker-independent system, the user's voice pattern is analyzed and compared to all other voice samples in the user database. The closest match to the particular voice data presented for identification becomes the identity of the speaker.

There are two variants of voice recognition—speaker-dependent and speaker-independent.

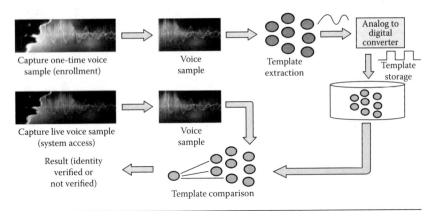

**Figure 1.6**   Voice recognition.

Speaker-dependent voice recognition relies on the knowledge of candidate's particular voice characteristics. This system learns those characteristics through voice training (or enrollment).

- The system needs to be trained on the users to accustom it to a particular accent and tone before employing to recognize what was said.
- It is a good option if there is only one user going to use the system.

Speaker-independent systems are able to recognize the speech from different users by restricting the contexts of the speech such as words and phrases. These systems are used for automated telephone interfaces.

- They do not require training the system on each individual user
- They are a good choice to be used by different individuals where it is not required to recognize each candidate's speech characteristics

*1.5.4.1 Differences between Voice and Speech Recognition* Speaker recognition and speech recognition are mistakenly taken as same; but they are different technologies.

| SPEAKER RECOGNITION (VOICE RECOGNITION) | SPEECH RECOGNITION |
|---|---|
| The objective of voice recognition is to recognize WHO is speaking. | The speech recognition aims at understanding and comprehending WHAT is spoken. |
| It is used to identify a person by analyzing its tone, voice pitch, and accent. | It is used in hand-free computing, map, or menu navigation. |

*1.5.4.2 Merits of Voice Recognition*
- It is easy to implement.

*1.5.4.3 Demerits of Voice Recognition*
- It is susceptible to quality of microphone and noise.
- The inability to control the factors affecting the input system can significantly decrease performance.
- Some speaker verification systems are also susceptible to spoofing attacks through recorded voice.

*1.5.4.4 Applications of Voice Recognition*
- Performing telephone and Internet transactions
- Working with interactive voice response (IVR)-based banking and health systems
- Applying audio signatures for digital documents
- In entertainment and emergency services
- In online education systems

## 1.6 Behavioral Biometrics Data Acquisition Device

The performance of behavioral biometrics critically depends on the quality of biometric data [13]. Sensor design and deployment requires high quality data for improved accuracy and flexible acquisition of data with high user acceptability.

Data acquisition devices for signature recognition are classified into two categories—online and offline. Scanners are generally considered as offline input device. It is a device that captures images from photographic prints, posters, and similar sources for editing and display. Flat bed scanners are popularly used to convert images and text into a digital format. Data acquisition is performed in offline mode after completion of writing procedure. Its nature is considered as static.

Graphic tablets are generally used for online signature acquisition. It allows to draw on the tablet in a natural way (just like pencil and paper) and their drawings appear on the screen of computing device. Its nature is generally considered as dynamic. These generate electronic signals representative of the signature during writing procedure. The generated signals can be coordinate signals, pressure and force signals, pen-down (operation of pulling down the tip of the pen toward the writing plane), and pen-up (operation of lifting up the tip of the pen from the writing plane) signals.

Keystroke dynamics acquires data through computer keyboard. A keyboard has characters printed on the keys and each press of a key typically corresponds to a single written symbol. However, to produce some symbols requires pressing and holding several keys simultaneously or in sequence. Most of the keys of keyboard produce letters, numbers, or signs, other keys or simultaneous key presses can produce actions or execute computer commands. In normal usage, the keyboard is used as a text-entry interface to type text and numbers into a word processor, text editor, or other programs. The interpretation of key presses is generally left to the software. A computer keyboard distinguishes each physical key from every other keys and reports all key presses to the controlling software. A command-line interface is a type of user interface operated entirely through a keyboard.

Mouse dynamics acquires data through mouse, which is a pointing device used on the screen of a computer. It enables user to execute commands or issue instructions to the computer by controlling a pointer on the screen. A mouse typically controls the motion of a pointer in two dimensions in a GUI. The mouse turns movements of the hand backward and forward, left and right into equivalent electronic signals. The relative movements of the mouse on the surface are applied to the position of the pointer on the screen, which signals the point where actions of the user take place.

Touch data are collected by touch screen input device. A user can give input or control the information processing system through simple or multitouch gestures by touching the screen with a special stylus or pen or one or more fingers. Some touch screens use a special stylus or pen only. The user can use the touch screen to react to what is displayed and to control how it is displayed. As for example, zooming is used to increase the text size. The touch screen enables the user to interact directly with what is displayed, rather than using a mouse, touchpad, or any other intermediate device.

Modern mobile devices come with various sensors such as accelerometer, gyroscope, GPS receiver, WiFi receiver, and so on. Few of the widely utilized sensors are mentioned below to present how to retrieve sensor data systemically.

An accelerometer is a device that measures proper acceleration experienced by an object relative to a free-falling frame of reference. A triaxial accelerometer installed on a mobile or wearable device returns a real-valued estimate of acceleration along the $x$, $y$, and $z$ axes in units of meter per second squared (m/s$^2$). By measuring the amount of static acceleration due to gravity, it is possible to find out the angle the device is tilted with respect to the earth. By sensing the amount of dynamic acceleration, we can analyze the way the device is moving. Because of the information that the accelerometer can offer, it can be employed as a high-bandwidth side channel to learn certain behavioral patterns.

A gyroscope is a device for measuring or maintaining orientation, based on the principles of angular momentum. It is also known as angular rate sensors or angular velocity sensors. Gyroscopes can sense the angular velocity along the $x$, $y$, and $z$ axes of a mobile or wearable device, corresponding to pitch, roll, and yaw, respectively, in units of radian per second (rad/s). The introduction of gyroscopes into the mobile devices has allowed for more accurate recognition of movements within 3D space than lone accelerometer devices. That is why modern smartphones are usually equipped with both accelerometer and gyroscope, for example, HTC, Nexus, iPhone, Nokia, and so on. Gyroscopes play a significant role in the gaming arena by providing a richer experience in handling the game.

Gait recognition acquires data through video camera that is used for electronic motion picture acquisition. Video cameras are used primarily in two modes. In one mode, the camera feeds real-time images directly to a screen for immediate observation. Most of the live connections are for security, military, and industrial operations where remote viewing is required. In another mode, the images are recorded to a storage device for archiving or further processing. Recorded video is used in surveillance and monitoring tasks in which unattended recording of a situation is required for later analysis. Closed-circuit television (CCTV) generally uses pan–tilt–zoom cameras (PTZ), for security, surveillance, and monitoring purposes. Such cameras are designed to be small, easily hidden, and able to operate unattended. These are often meant for use in environments that are normally inaccessible or uncomfortable for humans.

Voice and speech recognition system acquires data through microphone. A microphone is a transducer that converts sounds into variation of voltage. Electromagnetic transducers facilitate the conversion of acoustic signals into electrical signals from air-pressure variation. Microphones typically need to be connected to a preamplifier before the signal can be amplified with an audio power amplifier and a speaker. Microphone connects to software to convert human speech into commands or text.

It provides a natural interface with computing device that allows for new users to execute commands without having to learn complex command set. Speech recognition might not recognize the difference between similar words that is "their" and "they're" or be able to understand regional accents.

## 1.7 Behavioral Biometrics Recognition Systems

Behavioral biometrics is a science that deals with human identity on the basis of our personal features, characteristics, and patterns which we furnish. Behavioral biometrics recognition system is a type of pattern recognition system. A general description of the pattern recognition system is given here.

### 1.7.1 Accomplishment of Behavioral Biometrics Systems

Behavioral biometrics system mainly consists of the blocks given in Figure 1.7.

In Figure 1.7, universal object recognition system is presented. Two important aspects in object recognition are analysis and processing of objects. In order to understand the whole system, basic information is given here about initial processing and analysis of biometric traits as an essential step for object classification for recognition.

### 1.7.2 Initial Processing and Analysis of Biometric Traits

The results of object analysis with and without initial processing (preprocessing) are shown in Figure 1.8. During object analysis, it should be checked whether objects are free of noise or not. This is achieved during initial transformation of the original object, usually called preprocessing. It can be considered as an iterative process.

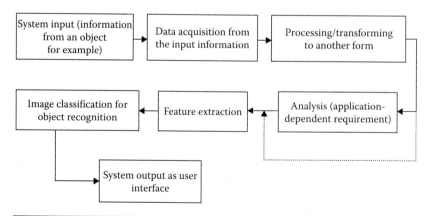

**Figure 1.7**  Universal object recognition system. (From K. Saeed and T. Nagashima [eds.], *Biometrics and Kansei Engineering*, Springer, New York, 2012. With permission.)

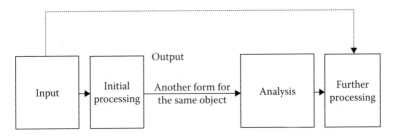

**Figure 1.8**   Processing and analysis. (Data from K. Saeed and T. Nagashima [eds.], *Biometrics and Kansei Engineering*, Springer, New York, 2012. With permission.)

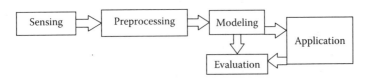

**Figure 1.9**   Behavioral biometrics framework.

*1.7.3 Framework*

A generic framework of behavioral biometrics is presented here (Figure 1.9). It has five major components.

1. *Sensing*: In order to build a behavior model for a behavior-aware application, required data are first collected as input to the framework. The data may come from hardware sensors embedded in smart devices, for example, GPS, gyroscope, and so on, or from device input/output interface.
2. *Preprocessing*: This component preprocesses the raw sensor data or human–device interactions from various sensor inputs. Selecting the right set of features is important for improving the target system performance and reducing overhead.
3. *Modeling*: Behavior modeling is the most challenging part. Given the constructed feature set, various modeling tools may be applied.
4. *Applications*: There are many use scenarios that can benefit from study of behavioral biometrics. Typical use-cases include anomaly detection, user classification, and user behavior prediction.
5. *Evaluation*: Performance of each application can be evaluated based on behavior learned from suitable models.

Sensing is the prerequisite for other components. Without sensory data collected from user devices, it is impossible to model user behavior and to realize applications. The actual modeling involves data preprocessing and model development. The application also depends on developed model.

## 1.8  Generalized Algorithm

In this section, authors describe a generalized algorithm for behavioral biometrics, which can be applied to any type of human activity.

The first step is to break up the behavioral trait into a number of atomic operations, each one corresponding to a single decision. Ideally all possible operations should be considered. In some cases, a large subset of most frequent operations may be sufficient.

User's behavior should be observed to produce frequency count for the occurrence of the atomic operations. The resulting frequency counts form a feature vector, which is used to verify the user based on the similarity score produced. An experimentally determined threshold value serves as a boundary for taking final decision by separating legitimate users from intruders. In case if user identification is attempted, a neural network or a decision tree approach might be used to select the best matching user from the database of existing templates.

The outline of the generalized algorithm is presented below.

Step 1: Pick behavior
Step 2: Break-up behavior into component actions
Step 3: Determine frequencies of component actions for each user
Step 4: Combine results into a feature vector profile
Step 5: Apply similarity measure function to the stored template and current behavior
Step 6: Experimentally determine a threshold value
Step 7: Verify user based on the similarity score comparison to the threshold value

Step 5 is not trivial. Several researches exist to determine what makes a good similarity measure function for different biometric systems. A good similarity measure takes into account statistical characteristics of the data distribution assuming enough data are available to determine such properties. Alternatively, expert knowledge about the data can be used to optimize a similarity measure function, for example, a weighted Euclidian distance function can be developed if it is known that certain features are more valuable than others. The distance score has to be very small for two feature vectors belonging to the same individuals (intra-class variability). At the same time it needs to be as large as possible for feature vectors coming from different individuals (inter-class variability).

## 1.9 Performance Measurement

Behavioral biometric systems are specific. Many characteristics make them difficult to define and to quantify their performance.

The biometric template contains generally temporal information. As for example, for keystroke dynamics analysis, we generally use a template composed of a set $N$ value couples $\{(Di, Fi) \; i = 1...N\}$ where $N$ is the number of characters in the password, Di is the duration time the user presses a key, and Fi is the time between this key and the next one in the password typing. For voice recognition systems, the biometric template is a sampled signal. Thus, the biometric template is generally quite important in size meaning that the parameters space is high.

The biometric template can change with time according to users. If we keep in mind the example of keystroke dynamics analysis, users with time learn how to type more efficiently their password. That means that the biometric template can be quite different compared to the one obtained after the enrollment step. Another example, the dynamics of signature can also change a lot with time as it becomes a reflex for the user. This variability has multiple consequences. The first one concerns the number of templates for the enrollment step that is generally higher than other types of biometric systems. The second consequence is that the verification/identification algorithm should take into account this variability in order to make a correct decision.

Another point concerns the testing of such biometric systems with biometric data that must embed this difficulty. The behavior as biometric characteristic can be very different for an individual given its age, culture, and experience. The evaluation of a behavioral biometric system is often realized considering a large diversity of users.

### 1.9.1 Benchmark Definition

Benchmark definition is really important for the performance evaluation of biometric systems. A benchmark database can be composed of real biometric templates (from test users) or synthetic ones. The definition of synthetic templates is easier for behavioral biometric data. Indeed, many behavioral modalities can be synthesized rather easily such as keystroke dynamics, voice, lip movements, mouse dynamics, signature dynamics, and so on. For morphological biometric modalities, it is much more difficult to do. The ability to generate more easily synthetic biometric templates is an advantage for the evaluation of such systems. Generally, a biometric model (generated after the enrollment phase) is computed for the same person given 2 or 3 capture sessions. The difficulty of behavioral biometric systems is that the biometric template naturally changes with time. Indeed, a human is a nice machine who wants to do things quicker. As a consequence, a benchmark database for behavioral modalities needs more capture sessions in order to take into account this variation. The number of capture sessions is important but also the period of time between them. This shows the difficulty and the cost of such benchmark definition for this type of biometric modality. The variability of behavioral biometric templates is really important if we compare morphological ones. Indeed, the fingerprint of individuals from different cultures or age is not so different. If we consider now the behavioral biometric modalities such as the keystroke dynamics, voice, or gait, the associated template can be very different from individuals at different ages. As a consequence, the benchmark database must embed all the variability of biometric templates to be representative of real applications.

### 1.9.2 Robustness Analysis

The behavior of an individual is very dependent on many factors like his mood, emotion, tiredness, or health. As for example, voice recognition systems are very sensitive to all these factors. In order to be used in a real context, one has to test the robustness of a biometric system face to all these modifications. Behavioral biometric systems can be very sensitive according to the sensor (keystroke dynamics, mouse dynamics, etc.). The behavior of an individual can be different for multiple sensors. Indeed, the performance in terms of EER can be quite high (>10%) in this case. Another point concerns the robustness of behavioral biometric systems face to attacks. The main difficulty for these systems is that anybody can try to duplicate a biometric template. As for example, it is not very hard to launch the verification given its keystroke dynamics, voice, or gait. That does not mean that the chance to be authenticated is necessarily higher but it is very easy to make a try. For morphological biometric systems, it is much more difficult even if as for example, fingerprints can be duplicated with some effort to launch one verification.

### 1.9.3 Discussion

In order to evaluate a behavioral biometric system, one could use the general methodology described in the previous section by using benchmark databases and classical performance metrics. Nevertheless, several aspects must be taken into account to consider the specificity of such systems:

- The number of sessions for the definition of biometric templates for testing such a biometric system must be high ($\geq 3$).
- The sessions must be relatively spaced in order to take into account the natural change of behaviors of individuals.
- Behavioral biometric templates can be in general easily synthesized. This approach for the definition of a benchmark database is interesting. It allows to test a large number of biometric templates and to control their alterations to quantify the robustness of the system.
- The benchmark database must contain some fake biometric templates to also test the robustness of the system.
  - A benchmark database must embed a large diversity of users (culture, age, etc.).
  - The performance evaluation of behavioral biometric systems must be realized using the same sensor during the enrollment and verification/identification steps.

## 1.10 Evaluation of Behavioral Biometric Systems

Performance evaluation is very much important on the efficiency of behavioral biometric systems. The metrics frequently used to quantify the performance of a behavioral biometric system are [11]

- *Computation time*: Necessary time required for the acquirement, enrollment, and verification/identification.
- *True positive (TP)*: Number of users that have been correctly authenticated.
- *False positive (FP)*: Number of impostors that have been authenticated.
- *FRR*: Proportion of authentic users that are incorrectly denied. It is calculated as

  FRR = 1 − TP/(number of genuine users)
- *FAR*: Proportion of impostors that are accepted by the biometric system. It is calculated as

  FAR = FP/(number of impostor users)
- *Failure-to-enroll (FTE) rate*: Proportion of user population for whom the biometric system fails to capture or extract usable information from biometric sample. This failure may be caused due to behavioral or physical conditions pertaining to the subject which hinder its ability to present correctly the required biometric information.
- *Failure-to-acquire (FTA) rate*: Proportion of verification or identification attempts for which a biometric system is unable to capture a sample or locate an image or signal of sufficient quality.
- *False match rate (FMR)*: The rate for incorrect positive matches.
- *False nonmatch rate (FNMR)*: The rate for incorrect negative matches.
- *Receiver operating characteristic (ROC) curve*: The method most commonly used to assess the performance of a behavioral biometric system is the ROC curve. The aim is to plot a curve representing FAR according to the FRR. In order to plot this type of curve, the value of the decision thresholds need to be changed. For each value of the threshold, associated FRR and FAR are calculated before plotting on the curve. In order to compare several biometric systems, the area under the curve

(AUC) is computed to determine EER where FAR = FRR. The optimal result is obtained if the AUC is equal to 1 and the EER is equal to 0.

- *Detection error trade-off* (DET) *curve*: DET curve is a ROC curve which has its linear scale replaced by a scale based on a normal distribution, to make it more readable and usable. In this case, the curve flattens and tends toward the right. The benefits of the DET curves are the same as those of ROC curves. In addition, they allow comparison among biometric systems having similar performance.
- *Cumulative match characteristic (CMC) curve*: This curve plots the identification rank values on the $x$-axis and the probability of correct identification at or below that rank on the $y$-axis.
- *Precision/recall (PR) curve*: This curve has a similar behavior to ROC curves. In order to draw the PR curve, we plot the positive predictive value (PPR = TP/(TP + FP)), also known as the precision versus the recall. We can then compute AUC and EER in a similar way as in ROC curves. One advantage is that we do not need the number of true negative in this method.

### 1.10.1 Discussion

In order to evaluate a behavioral biometric system, one could use the general methodology by using benchmark databases and frequently used performance metrics. Several aspects must be taken into account to consider the specificity of such systems.

- Number of sessions for the definition of biometric templates for testing such a biometric system must be high.
- Sessions must be relatively spaced in order to take into account the natural change of behaviors of individuals.
- Behavioral biometric templates can be in general easily synthesized. This approach for the definition of a benchmark database is interesting. It allows to test a large number of biometric templates and to control their alterations to quantify the robustness of the system.

- Benchmark database must contain some fake biometric templates to test the robustness of the system.
- A benchmark database must include a large diversity of users.
- Performance evaluation of behavioral biometric systems must be realized using the same data acquisition device during the enrolment and verification/identification steps.

## 1.11 Comparison and Analysis

All of the behavioral biometrics share a number of characteristics; so that, it can be analyzed as a good biometrics [11].

- *Universality*: Behavioral biometrics is dependent on specific abilities possessed by different people to a different degree or not at all and so, in a general population, universality of behavioral biometrics is very low. But since behavioral biometrics is only applied in a specific domain, the actual universality of behavioral biometrics is a 100%.
- *Uniqueness*: Since only a small set of different approaches to performing any task exist, uniqueness of behavioral biometrics is relatively low. Number of existing writing styles and varying preferences are only sufficient for user verification not identification unless the set of users is extremely small.
- *Permanence*: Behavioral biometrics exhibit a low degree of permanence as they measure behavior which changes with time as person learns advanced techniques and faster ways of accomplishing tasks. However, this problem of concept is addressed in the behavior-based intrusion detection research and systems are developed capable of adjusting to the changing behavior of the users.
- *Collectability*: Collecting behavioral biometrics is relatively easy and unobtrusive to the user. In some instances, the user may not even be aware that data collection is taking place. The process of data collection is fully automated and is very low cost.
- *Performance*: The identification accuracy of most behavioral biometrics is low particularly as the number of users in the database becomes large. However, verification accuracy is very good for some behavioral biometrics.

**Table 1.1** Comparative Analysis

| BEHAVIORAL BIOMETRICS TRAITS | DETECTION RATE (%) | FAR (%) | FRR (%) | EER (%) |
|---|---|---|---|---|
| Signature verification | | 1.6 | 2.8 | |
| | 95.7 | | | |
| Keystroke dynamics | | 0.01 | 4 | |
| Gait/stride | 90 | | | |
| Voice/speech | | | | 0.28 |

*Source:* Standring S (ed.), *Gray's Anatomy: The Anatomical Basis of Medicine and Surgery*, 39th edn., Churchill-Livingstone, New York, 2004. With permission.

- *Acceptability*: Since behavioral biometrics can be collected without user participation, they enjoy a high degree of acceptability, but might be objected to for ethical or privacy reasons.
- *Circumvention*: It is relatively difficult to get around behavioral biometric systems as it requires intimate knowledge of someone else's behavior, but once such knowledge is available, fabrication might be very straightforward. This is why it is extremely important to keep the collected behavioral profiles securely encrypted.

All behavioral biometrics essentially measure human actions specific to every human skills, style, preference, knowledge, motor skills, or strategy. While many behavioral biometrics are still in their infancy, some very promising research has already been done. The results obtained justify feasibility of using behavior for verification of individuals and further research in this direction is likely to improve accuracy of such systems.

Table 1.1 represents recognition, verification, and error rates of few behavioral biometrics traits.

## 1.12 Human Measurement and Evaluation on the Basis of Behavioral Biometric Features

Behavioral biometrics serves for human feature measurement [1]. However, the measurement has only one of two purposes (we call them modes), either to identify people (to answer the question "Who are you") or to verify them (to answer the question "Are you really who you claimed?").

### 1.12.1 Verification and Identification

To verify people, we will need to answer with either "Yes" or "No." It is a matter of binary classification. This kind of evaluation is actually easier than that concerning identification. To identify people, we need to seek database to evaluate the personality of someone and find out who he is.

To prevent more than one person from using the same identity, positive recognition is considered. Here we should follow the verification mode. However, when the goal of a biometric system is to prevent one user from making use of more than one identity, then we are dealing with negative recognition and hence identification mode is in use.

### 1.12.2 Error Sources in Behavioral Biometrics

There are some facts that should be known and considered while considering errors in behavioral biometrics domain [1]. There still exists a possibility that the biometric system may make some errors while verifying human features for their recognition.

This happens when we register user's features under different conditions of data acquisition (different sensors for data collection) or when they are collected under different conditions like lighting, atmospheric conditions like temperature, image resolution, age, and so on. This leads to the result that for one person we may have two different features, for example two characteristic signatures, voices, which are on contrary to the basic assumption that biometric features are unique. Consequently, there is a probability that our features match those of other people leading to some matching errors. Such situations may appear after poor capture and hence we have an error introduced by the device. Or, that the methodology we are following has some drawbacks leading to some errors. However, here we are dealing with errors that result from the human when interacting with biometric data acquisition devices: how they give their biometric features, the way they look at the camera, for example. Errors may also arise when the individual is sick or under treatment, or when they provide their data, for example their photographs, they supply old ones. This, sometimes, is completely different from their actual look.

## 1.13 Types of Basic Verification Errors and Their Rates

Based on the graphical representation of genuine and impostor distribution [1], a biometric verification system can make two types of errors:

1. Matching the recognition features of person A as belonging to person B, that is, an acceptance occurs, although the user is an impostor. This means the two users will be treated as one person leading to what is called false acceptance (FA) or false matching). The resulting error is then called false acceptance error, measured as a ratio or rate, and hence the name false acceptance ratio/rate (FAR).

2. Mismatching features of person A and treating them as belonging to another person. That is, a rejection occurs although the user is genuine. This in turn leads to what is called false rejection—FR or false nonmatching) and the error rate is called false rejection ratio—FRR.

Both FA and FR rates depend on the input source of information and the percentage of the noise.

### 1.13.1 Error Graphical Representation

In order to evaluate a biometric system for a degree of success in human recognition on the basis of their biometric features [1], errors should be either eliminated or at least decreased to a minimum. The more we can do is to eliminate such errors the better and more effective will be our system as it would bring lower mismatching probabilities and hence the rate of the recognition success is higher. Since there is always a possibility of system deviation from the ideal case (normal desired accuracy state), the mismatching (in both, positive and negative) error can be represented by a Gaussian curve. This helps to study the probability of system errors. Both genuine distribution (also called true positive) and impostor distribution (true negative) are drawn on one plane. Their overlapping shows some common areas to define the already defined error rates FAR and FRR. A selected error score value that defines the system tolerance or sensitivity is called threshold ($T$ in Figure 1.10). $T$ is fixed according to the application and a vertical line is drawn at $T$ to define these possible basic errors.

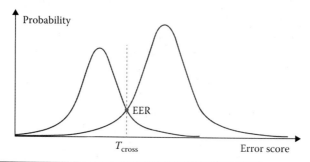

**Figure 1.10**  Gaussian distribution for the probability of error to explain FAR and FRR errors as overlapping areas between the genuine (registered user) and impostor (unregistered user) curves. (Data from K. Saeed and T. Nagashima [eds.], *Biometrics and Kansei Engineering*, Springer, New York, 2012. With permission.)

Obviously, FAR and FRR are both threshold dependent. Thus, if the threshold is selected to be at the cross point of the two curves ($T_{cross}$ in Figure 1.10), then at this point the probability of the two error rates are equal, FAR = FRR and the error rate is then called crossover error rate (*CER*) EER.

Another way of illustrating the error rates FAR and FRR is the direct relation between them, which is shown in Figure 1.11.

Increasing the threshold $T$ will result in a higher value of FRR, while decreasing $T$ would cause an increase in FAR. On the other hand, the position of $T$-line, that is, the value of the threshold for a given system is actually application-dependent. We cannot assign, for example, a lower value for $T$ to decrease the FRR (and hence, to increase FAR) and make the system more tolerant to system inputs

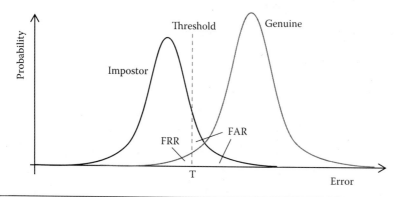

**Figure 1.11**  Equal error rate illustration. (Data from K. Saeed and T. Nagashima [eds.], *Biometrics and Kansei Engineering*, Springer, New York, 2012. With permission.)

just to obtain a system of higher success rate. The reason is simply we would obtain a nonsecure system at the time when the matter requires high security conditions. Therefore, in security applications, the FRR is not as important as the FAR. FAR should be very low (in some applications may reach 0.00001%) at a bit higher FRR (e.g., 0.01%), so that only authorized people can be given an access. However, in forensic applications the tolerance should be very low and hence FRR should be at the lowest possible levels (the closest to zero) and obviously, the system should not miss any single person suspected of committing a crime. In commercial applications, banking systems for example, $T$ is selected around what is called EER (*equal error ratio*) where FAR = FRR. Here, the time of recognition procedure should also be taken into consideration; it is usually about 1–4 s depending on the commercial application (Figure 1.12).

### 1.13.2 Further Study of Errors

We should first recall that in many normal cases, the environmental conditions and the sensor nature and accuracy play a role in creating new errors. The impact of such errors on the classification results is large and should always be considered. Moreover, it is also possible to depict the system performance for all threshold values. This is obtained by ROC curve, which plots FAR against (1 – FRR) or FRR for various values of threshold $T$ in ROC space.

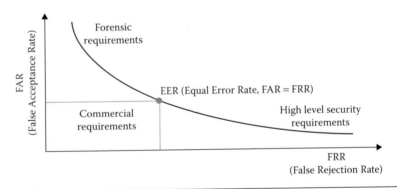

**Figure 1.12** FAR is high (forensic and criminal applications) for low FRR while FRR should be very high (in high security requirements), although FAR is very low then. EER is the most practical tolerance for commercial use (in banking, for example). (Data from K. Saeed and T. Nagashima [eds.], *Biometrics and Kansei Engineering*, Springer, New York, 2012. With permission.)

## 1.14  Open Issues

### 1.14.1  Collection of Sensitive Information

Some of the techniques use collected data which may be used to derive highly sensitive information. Perhaps the clearest example of this is the collection and analysis of EEG patterns. EEG patterns may be used simply for identification, but it is now possible to extract significant information about the thoughts of the subject from records of brain activity. Even less obviously, sensitive types of behavioral biometrics may incidentally gather highly sensitive information however. For example, gait features may reveal emotional features such as depression. It should be emphasized however that in some cases, such as keystroke dynamics, this issue is less important since there is less likely to be a correlation between the biometric measurement and more sensitive features of a person's profile (such as mood, health state, etc.). The use-case of the technique also affects the importance of this issue. For example, when used to detect intrusion into a home computer (which is assumed to be trusted), such sensitive data may be kept private to the user. When used to authenticate to a corporate web application, this may not be the case. Furthermore, techniques exist to limit the exposure from collected behavioral profiles, provided that the data collector is trusted (e.g., hashable profiles, etc.). Currently this area of research is in its infancy and will undoubtedly grow as does the adaptation of behavioral biometrics.

### 1.14.2  Negative Reaction to Obtrusive Equipment

Due to the above possibilities, as well as the obtrusiveness of the equipment used, there is a strong negative reaction on the part of end-users to certain kinds of biometrics, including behavioral biometrics such as EEG scanning. This is a significant problem for deployment.

### 1.14.3  Consent and Secondary Use for Data Collected with Unobtrusive Equipment

Some behavioral biometrics are at the other end of the obtrusiveness spectrum. For example, gait recognition and keystroke dynamics are possible without any impact on the subject. This has its own drawbacks, because it means that a subject may be identified without their awareness. Highly unobtrusive techniques offer the possibility of identifying people outside the scope of the original application, without any special equipment.

### 1.14.4 Sensitivity to Change of Application Configuration

Most of the behavioral biometrics are sensitive to the implementation context.

### 1.14.5 Spoofing Attacks

Coercive impersonation is a type of attack in which the attacker physically forces a genuine user to identify himself to an authentication system or removes the biometric (e.g., a finger) from the person to use as a key to gain access to the resources. Replay attack is based on recording a previously produced biometric such as taking a picture of a face or recording a person's voice and submitting it to the biometric data collection unit. Impersonation attack involves an attacker who can change his appearance to match a legitimate user for example use makeup to copy somebody's face or impersonate voice or forge a signature. Approaches to spoofing behavioral biometrics are similar to those for physical biometrics but with some domain-specific features. Replay attacks are very popular since it is easy to record an individual's voice or copy a signature. Human mimicking or forgery is also a very powerful technique with experts consistently breaching security of signature-based or voice-based authentication systems. On the other hand, the removal of body parts such as fingers or DNA, is less likely to be a useful attack. Just as software can be trained to recognize samples based on feature recognition, it can also be trained to generate samples containing a given set of features. Such techniques produce models of behavior parameterized with observed target user data which steadily improve in their performance. It may also be possible to perform a brute force attack on a biometric system by going through a large number of possible behavioral parameters.

## 1.15 Future Trends

The constraints and the cost for the evaluation of behavioral biometric systems are extremely prohibitive [11]. High quality benchmark databases must be available for the research community taking into account the previous constraints. These databases would facilitate the testing and the development of new behavioral biometric systems. They also would be able to compare different enrollment and identification/verification algorithms to increase the knowledge in the domain. It is generally difficult to say if the database is representative of real use-cases and if the system achieves better than others in the state of the art. The statistical evaluation of behavioral biometric systems is important but is not sufficient. A behavioral biometric system to be used in real conditions must be easy to use. Subjective evaluation is a domain that needs a lot of research to take into account the user as the central element in the behavioral biometric system.

### 1.16 Application Area

Behavioral biometric applications can be categorized in horizontal categories as well as vertical markets.

- *Citizen identification*: Identify/authentify citizens interacting with government agencies.
- *PC/network access*: Secure access to PCs, networks, and other computer resources.
- *Physical access/time and attendance*: Secure access to a given area at a given time.
- *Surveillance and screening*: Identify/authentify individuals present in a given location.
- *Retail/ATM/point of sale*: Provide identification/authentication for in-person transactions for goods/services.
- *E-commerce/telephony*: Provide identification/authentication for remote transactions for goods/services.
- *Criminal identification*: Identify/verify individuals in law enforcement applications.

In each of those applications, biometric systems can be used to either replace or complement existing authentication methods.

### 1.17 Behavioral Biometrics Used in Real-Time Application

Use of behavioral biometrics (HCI) is studied for real-time remote applications like remote health [12]. Collection of data run for a few months, so that sufficient amounts of transactions could be seen for the majority of the users. Once sufficient amounts of data are reached, analysis is performed on it to calculate the accuracy of the system.

After a user is verified with traditional security techniques, such as passwords, it can enhance the protection even after the user has logged in. It can continuously monitor the user during the whole working session to create a continuous authentication process. Verification module checks whether a user is accepted into a system or not. If a user who should be rejected is accepted, it results into FA. If a user that should be accepted is rejected, it results into FR. HCI uses a set of behavioral traits to calculate a similarity ratio between the current user's behavior and the expected. The similarity is combined with a threshold, so that if the similarity drops below the threshold, the user will be detected as an impostor. The score represents how similar a sample and a template are; the higher score the more similar they are. The threshold is a value that says that all samples that have a score above this value are considered to be originated from the correct user and below to be originated by an impostor.

The matching algorithm performs a decision based on a threshold, which determines how close to a template the input sample needs to be for it to be considered a match. If the threshold is reduced, there will be less false nonmatches, but more false accepts.

HCI uses evaluator to process user data captured by remote application interface. The profiles are built by chronologically inserting every transaction with the user. Here, every transaction made for each user is considered to have been made by the correct user. The scores received for each such insertion are stored and later used to calculate the FRR of the system in training phase.

By varying the threshold starting from 0 and going up to 100 by steps of 1 the FAR and FRR curves are calculated. FRR values are calculated by checking the amount of real transactions that are rejected for the given threshold.

Here simulated attacks are built up by combining data from other users, not necessarily reflecting a real attacker. Since the data comes from a live environment there could be attacks or account sharing in it which potentially could push the FRR to the worse.

Adjusting the threshold makes it possible to adjust the system to any specific needs. It can be adjusted for higher security but that would increase the risk for false alarms and vice versa. A lower threshold can be utilized if it is desirable that the correct user would always be accepted and successfully blocking fewer impostor attempts. If the threshold is lowered to achieve practically 0% false rejects, it is still possible to successfully block most impostor attempts.

## 1.18 Conclusions

Behavioral biometrics is mainly well suited for verification of end-users interacting with any input devices like computers, cell phones, and so on. As the number of electronic appliances used in homes and offices increases, behavioral biometrics become a promising technology. Future research should be directed at increasing overall accuracy of such systems, for example, by looking into possibility of developing multimodal behavioral biometrics; as people often engage in multiple behaviors at the same time, for example, talking on a cell phone while driving or using keyboard and mouse at the same time. Intensive works are going on [1] for removing the possibly existing drawbacks or undesired features. This would definitely result in improving the algorithmic level of performance and their operation time for faster, productive action and universal applications.

# References

1. Saeed K, Nagashima T (eds.) (2012) *Biometrics and Kansei Engineering.* Springer, New York.
2. Boulgouris NV, Plataniotis KN, Micheli-Tzanakou E (eds.) (2009) *Biometrics: Theory, Methods, and Applications.* John Wiley & Sons, Hoboken, NJ.
3. Bolle RM, Connell JH, Pankanti S, Ratha NK, Senior AW (2010) *Guide to Biometrics.* Springer, New York.
4. Rybnik M, Panasiuk P, Saeed K, Rogowski M (2012). Advances in the keystroke dynamics: The practical impact of database quality. *Computer Information Systems and Industrial Management*, Volume 7564 of the series Lecture Notes in Computer Science, pp. 203–214.
5. Nishiuchi N, Yamaguchi T (2008) Spoofing issues and anti-spoofing. *International Journal of Biometrics*, 1(2), 365–370, Inderscience Publishers, UK.
6. Yampolskiy RV, Govindaraju V (2008) Behavioural biometrics: A survey and classification. *International Journal of Biometrics*, 1(1), 81–113.
7. Abdulla WH, Yu PWT, Calverly P (2009) Lips tracking biometric for speaker recognition. *International Journal of Biometrics*, 1(3), 288–306.

8. Nishiuchi N, Komatsu S, Yamanaka K (2010) Biometric verification using the motion of fingers: A combination of physical and behavioral biometrics. *International Journal of Biometrics*, 2(3), 222–235.

9. Molenberghs G (2006) Biometry and biometrics. *Sensor Review*, 26(1). http://www.emeraldinsight.com/journals.htm (accessed on May 28, 2012).

10. Saeed K (2011) A note on problems with biometrics methodologies. In *Proceedings of IEEE-ICBAKE Conference, IEEE CS Press – CD*, Takamatsu, Japan, September 19–21, pp. 20–22.

11. Adler A, Youmaran R, Loyka S (2006) Towards a measure of biometric information. http://www.sce.carleton.ca/faculty/adler/publications/2006/youmaran-ccece2006-biometricentropy.pdf (accessed on August 2, 2006).

12. Bhattasali T, Saeed K (2014) Two factor remote authentication in healthcare. In *Proceedings of International Conference on Advances in Computing, Communication and Informatics*, IEEE XPLORE, New Delhi.

13. Saeed K, Pejaś J, Mosdorf R (eds.) (2006) *Biometrics, Computer Security Systems and Artificial Intelligence Applications*. Springer Science + Business Media, New York.

14. Standring S (ed.) (2004) *Gray's Anatomy: The Anatomical Basis of Medicine and Surgery*, 39th edn., Churchill-Livingstone, New York.

# 2
# SIGNATURE RECOGNITION

Initial studies on handwriting can be attributed to the area known as graphology. The term "graphology" was first used by St. Jean-Hippolyte Michon in Paris in 1887, and comes from a combination of two Greek words: "graphein"—write, and "logos"—the science. However, the first attempts of systematic analysis of writings are contributed to Camillo Baldini whose work entitled "Treated how, by a letter missive, one recognizes the writer's nature and qualities" from seventeenth century is considered the starting point of today's handwriting analysis. The main interest of graphology is the interpretation of handwriting in order to determine personality traits of the writer and the assessment of his state of health (mental and physical). Currently, however, graphology applied to the psychological assessment is usually regarded as pseudoscience.

A handwritten signature is widely used in everyday life for authorization of various kinds of documents and for confirmation of financial operations (credit card payments, for example). Handwritten signature is also actively studied as behavioral trait in biometrics. There are many methods and commercial systems that are focused on automatic verification of handwritten signatures [1]. Despite the years of research devoted to this subject no perfect solutions were found and there is still field for improvements.

## 2.1 Brief History of Handwriting Analysis

The scientific methods applied to the analysis of handwriting can be found in the discipline called questioned document examination. The primary objective of this field is the identification of a person who created a piece of writing or a signature, which is an important aspect of many legal investigations where documents containing handwriting or typewritten text are part of evidentiary material. Typical documents that are analyzed by forensic document examiners include wills, contracts, bank checks, and threatening letters.

Analysis of the signature takes a special place in the field of questioned document examination. This is related to its extensive usage in formal procedures established by the letter of law—the assessment signature authenticity is an important part of many lawsuits and is crucial for their progress. The second reason is the unique nature of the signature, which distinguishes it from an ordinary handwriting and requires special techniques. Due to its size, the signatures contain a very small amount of information that can be analyzed. For comparison, fragments of handwritten texts such as letters or wills allow for a much more comprehensive assessment of the writer's style. Small amount of material which is the carrier of individual characteristics requires a more detailed analysis of the signature shape and utilizing additional information such as pressure and dynamics of writing that are considered crucial for proper assessment of genuineness.

## 2.2 Automated Systems for Signature Recognition

Studies on automatic systems for signature recognition started in early 1960s. One of the first complete system for handwritten signature recognition was developed by J. Marcel in 1965 and described as a technical report [2]. The solutions in this time focused solely on the analysis of geometric features of the signature shape and did not include information on dynamics [3]. This condition persisted until the late 1970s and was partly caused by limited possibilities of recording equipment. The analysis of dynamic features made its way into signature biometrics when first hardware devices became available that could register pen movements during writing. One of the first specialized devices that allowed to capture dynamic signature was proposed in patent US4035768 in 1977.

The style of handwriting and singing is specific to every individual and this property is utilized by the experts in the field of document examination. However, as biometric trait, it should be assessed using criteria similar to other biometrics that were proposed in [4]. Table 2.1 shows signature in comparison with other biometric traits using the following criteria: universality (availability of trait in population), distinctiveness (ability to distinguish individuals), permanence

**Table 2.1** Comparison of Signature with Other Biometric Traits

| BIOMETRIC IDENTIFIER | UNIVERSALITY | DISTINCTIVENESS | PERMANENCE | COLLECTABILITY | PERFORMANCE | ACCEPTABILITY | CIRCUMVENTION |
|---|---|---|---|---|---|---|---|
| *Signature* | L | L | L | H | L | H | H |
| Face | H | L | M | H | L | H | H |
| Fingerprint | M | H | H | M | H | M | M |
| Hand geometry | M | M | M | H | M | M | M |
| Keystrokes | L | L | L | M | L | M | M |
| Hand veins | M | M | M | M | M | M | L |
| Iris | H | H | H | M | H | L | L |
| Retinal scan | H | H | M | L | H | L | L |
| Voice | M | L | L | M | L | H | H |
| Facial thermogram | H | H | L | H | M | H | L |
| DNA | H | H | H | L | H | L | L |
| Gait | M | L | L | H | L | H | M |
| Ear | M | M | H | M | M | H | M |

*Source:* A.K. Jain et al., *IEEE Trans. Circuit Syst. Video Technol.,* 14, 4–20, 2004. With permission.

(invariance over period of time), collectability (possibility of quantitative measurement), performance (achievable recognition accuracy and speed, required resources), acceptability (acceptance of measuring and everyday usage), and circumvention (resistance to fraud). As can be seen from Table 2.1, a signature evaluated as a biometric trait has both advantages and disadvantages. Its main benefits are ease of registration and acceptance of everyday usage. On the other hand, it has many drawbacks: relatively low permanence—especially over longer periods of time (signature given by one person can vary significantly and these variations may increase over time), problems with signature uniqueness in larger populations (simple signatures given by different individuals may have similar shape and dynamic characteristics), ease of producing forgeries, and fouling the system (especially systems that use only static information are susceptible to fraud attempts). However, some of these drawbacks can be reduced by including dynamic features during signature verification. Ongoing research on biometric methods for signature recognition also resulted in lowering error rates to acceptable levels for many applications. Low permanence can be addressed by updating reference signatures with re-enrollment. Moreover, the ability to change a signature can be also seen as advantage in case of a forgery. In this case, the user can provide a new version of his signature that is unknown to impostor and use it in future authentications.

## 2.3 Offline and Online Signatures

Handwritten signatures are divided into two main categories:

- *Offline*: Is a static representation in a form of a signature image acquired by scanning a paper document or taking a picture with digital camera.
- *Online*: The acquisition is performed during act of signing and includes both the image and dynamics of writing.

Offline signature contains complete information on signature shape. The main drawback of using offline data is that it is very difficult or sometimes impossible to detect forgeries. Signature shape can be easily imitated when a forger has the access to examples of authentic signatures and when comparing only the shapes of original and forged

**Figure 2.1**  Examples of offline signatures extracted from paper forms.

samples, it may be very difficult to distinguish them. However, in many applications where handwritten signatures are given as an on-a-paper form, only static information in the form of a signature image is available. Figure 2.1 shows examples of offline signature images extracted from paper forms.

In case of online signatures the registration includes not only the shape of the signature but also the information that describes the way the signature was written. Basic dynamic data gathered at the time of signing usually contains the following parameters (Figure 2.2): $X$, $Y$-coordinates, pressure ($P$), elevation ($L$), and altitude ($A$). The coordinates $X$ and $Y$ determine the position of the pen tip inside the controlled area where the signing process is being traced. The pressure parameter describes the pen pressure inflicted on the tablet surface. The altitude is the angle between the pen and the surface. The azimuth

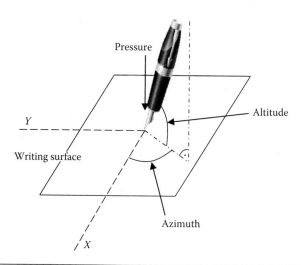

**Figure 2.2**  Online signature parameters registered during act of signing. (Data from M. Adamski et al., Signature system based on extended shape context descriptors, *International Conference on Biometrics and Kansei Engineering*, IEEE, Tokyo, Japan, pp. 267–272, 2013. With permission.)

denotes the angle between the projection of the pen onto the writing surface and the $X$-coordinate axis.

Dynamic information acquired during online registration is very important because it allows to increase the system resistance to forgeries. Imitation of dynamic changes of pen pressure, pen angles, and drawing trajectory is much more difficult than just copying a signature image. These dynamic parameters are also called hidden because it is impossible to precisely reconstruct their characteristics given only the image of a genuine signature. Another advantage when using online data is that it is much easier to analyze—there is no need to extract a signature from complex background or deal with artifacts resulted from poor quality of scans.

Data describing online signature takes the form of sequences of samples registered at a specific frequency. Examples of such sequences are shown in Figure 2.3. The values of samples can be treated as functions of time of registration. On the basis of these functions other

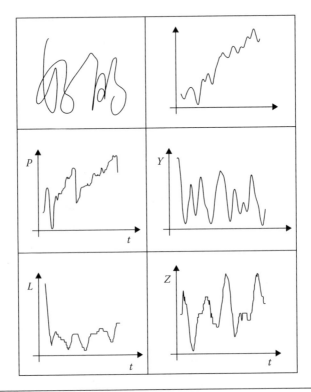

**Figure 2.3** Example of online signature and its characteristics.

characteristics can be computed such as first and second derivatives interpreted as pen velocity and acceleration.

## 2.4 Types of Forgeries

One of the most important properties of signature verification system is its resistance to forgery attempts. In order to asses this property the system is tested against forged signatures prepared for genuine signatures used for the examination. There are several types of forgeries that differ in degree of similarity to genuine samples and the technique that was used to create them. The following description defines forgery classes that are most commonly used in the biometric signature systems evaluation.

"Random forgeries" are created from original signatures belonging to other individuals—random imitation is original signature of person A is used as a forgery for of a signature of person B. This kind of forgeries can be easily detected because their shape has different geometrical properties from the shape of genuine samples.

"Simple forgeries" or "casual forgeries" are produced based on personal information such as first name and family name without any knowledge about the shape of genuine signature. Those imitations are usually written as a combination of the initials and the fully spelled family name. Their shape, similarly to random forgeries, is usually totally different from the shape of genuine signatures and therefore can be easily distinguished.

Random and simple forgeries are sometimes treated as the same class of forgeries due to the fact that they both lack geometrical resemblance to original samples [6].

"Skilled forgeries" are imitations created by forgers who have access to genuine examples and can spend as much time as required to train how to imitate original signature shape. This kind of forgery is more difficult to differentiate from genuine signatures because their shape is very similar to original. Some researchers introduce subcategories of this class: "traces" and "simulated forgeries" [7].

"Traced forgeries" are made by using carbon paper or other techniques that allow to replicate original signature by following its ink line. Such imitation is a precise copy of the original sample. "Simulated forgeries" are made by trained forgers who create signature imitations "from memory" based on original samples available to them beforehand. Quality of such imitations depend on forger skills, however, they are never exact copies of original samples.

In work [8] three other categories are presented: "blind forgeries"—produced based on descriptive or textual knowledge of original signatures without example of its actual shape, "low-force forgeries"—the impostor has access to genuine signatures and can use tracing or memorized shape to create an imitation, "brute-force forgeries"—mimic not only shape of the signatures but also dynamic features such as time of signing, order of and timing of drawing signature components. In order to create "brute-force forgery," the impostor must have access to information on dynamics of the signing process such as a video sequence or timed data captured with tablet device.

There is another classification present in graphometry and document analysis that is also used in some works on biometric systems [9]. It introduces for four categories: "zero-effort forgery," "home-improved forgery," "over-the-shoulder forgery," and "professional forgery." As the name suggest, "zero-effort forgery" is simplest and easiest to detect—the forger do not know the shape of the original signature and gives his own signature or writes any kind of shape. In the case of "home-improved imitation," the forger has a copy of original signature and can improve ("at home") his skills in imitating its shape. "Over-the-shoulder forgery" imitates the shape and dynamic characteristics of the original signature—the impostor can learn by observing the process of signing over the writer's shoulder. "Professional forgeries" are created by individuals who are experts in document examination and can use their knowledge to produce high quality imitations.

It may seem legitimate to say that to evaluate the signature recognition system one should use only the skilled forgeries. The authors of many works, however, especially in static systems, use only random and simple types. Using these simplest forms of forgeries may be justified by the fact that 95% of falsification of signatures in banks are imitations of these two types [6].

## 2.5 Databases for Signature System Evaluation

In order to examine the effectiveness of the biometric system a properly prepared database is needed, which contains examples of genuine and fake signatures. The development of such a database in-house requires considerable effort. Another difficulty when using your own database of examples is comparison of obtained the results with other systems that have been tested using a different set of signatures.

Preparing database for evaluation and comparison of methods requires careful development of the registration procedure. The procedure should consider the following aspects:

- *The effect of the device, pen, and writing surface*: The use of different types of tablets, pens, and writing surfaces affects the recorded signature, therefore the procedure should establish the type of tablet/pen/surface or several types if one wants to examine their impact on the signature characteristics.
- *The influence of the environment*: Conditions under which signature is given can also influence its shape and characteristics. In many everyday situations the signatures are given standing up, while in others the documents can be when sitting on a chair—the signatures provided in both cases can differ due to changed balance and body position. In order to create representative set of samples the registration process should include such cases or use single repeatable setting that mimics the environment in which the particular system will be used.
- *The effect of time interval*: Signatures submitted in the intervening period have larger differences than written one after the other, data collection should be carried out in sessions separated with predefined interval of time.

The source of offline signatures are scanned documents, such as contracts, forms, and financial documents. In these documents the signatures are placed in specific fields. To create offline signature database simple forms are usually used with a grid of rectangles in which individuals put their signatures. When preparing the forms size these rectangle fields should correspond to the size of such fields in typical documents.

Completed forms should be scanned with the appropriate resolution (no less than 300 DPI) to retain all important features. Data should not be written in lossy compression formats which may introduce artifacts that hinder the extraction of the signature line.

When registering online, attention should be paid to resolution and sampling frequency of the device. Studies indicate that the frequency components of handwriting do not exceed 20 Hz. According to the Nyquist theorem to register these components is sufficient to frequency above 40 Hz.

Using tablet differs from writing on a paper and may cause some difficulties—the signatures registered may differ from those written on a paper. Therefore, before the actual registration users should be given the opportunity to practice signing with the tablet device. An alternative is to use tablets that allow capturing data while writing on a piece of paper.

For research on biometric signature systems one of the publicly available databases can be used. The advantage of using them is the ability to compare against other systems that were tested using those databases. The following section introduces signature datasets described in the literature.

### 2.5.1 SVC2004

This database was initially prepared for signature verification competition that was carried out in 2004 [10]. It contains online signatures registered with WACOM Intuos tablet. Acquired features (registered at 100 Hz sampling rate) include $X$-coordinate (scaled), $Y$-coordinate (scaled), time stamp event, pen up/down status, azimuth, altitude, and pressure.

For every person 20 genuine signatures and 20 skilled forgeries are available. The signers were advised to not to use their true signatures for security reasons.

Signatures were collected in two separate sessions in the interval of one week. In the first session every person contributed 10 genuine signatures. Each signer could practice signing before actual registration and could examine acquired samples and reject them if they were not satisfying. During the second session each person provided additional 10 genuine samples and 20 forgeries (4 forgeries for 5 selected users). During forgery attempt users were provided with genuine samples and could replay the signing sequence on the computer screen. They could also practice before forgery acquisition. The database contains mostly English and Chinese signatures. The signature set that is publicly available contains signatures from 40 individuals giving a total of 1600 signatures ($40 \times 20$ genuine + $40 \times 20$ forgery).

### 2.5.2 GPDS-960

GPDS contains offline genuine and skilled forgery signatures of 960 persons [11]. For each person, the database contains 24 genuine signatures and 30 forgeries. Genuine signatures were collected in one session using a paper form. The form was divided into 24 boxes, with sizes 2.5 × 2.5 cm and 5 × 2.5 cm that are typical to official documents such as checks, credit card vouchers, and others.

The forgeries were created by 1920 individuals different from the group providing authentic signatures. They also used paper form with examples of authentic signatures and could devote as many time as needed to train forgery skills. Their form contained examples of five different genuine signatures selected at random, and three boxes below for providing three forgeries per genuine sample. As a result, each forger created 15 imitations (3 imitations per 5 users). Individuals who created forgeries were not experts so the forgeries cannot be treated as professional. The total amount of signatures available is 28824 (960 × 24 genuine + 960 × 30 forgery).

### 2.5.3 MCYT-100

MCYT was created in Spain as bimodal biometric database that contains fingerprints and signatures collected from 330 individuals [12]. MCYT-100 is part of the complete corpora that contains online signatures of 100 users and can be used for research. For every subject in the database 25 genuine and 25 skilled forgeries are available. The signatures were acquired with WACOM Intuos A6 tablet. Acquired features include $X$-coordinate, $Y$-coordinate, azimuth, altitude, and pressure and were captured with 100 Hz sampling rate. Each individual provided 25 genuine signatures and 25 forgeries using the following procedure: user $n$ writes his own signature 5 times, and then writes 5 forgeries of genuine signature of user $n - 1$. Before writing these forgeries, a static image of genuine signature of user $n - 1$ is provided to user $n$ so he can exercise coping it (at least 10 times) before producing 5 imitations that are registered. This procedure is repeated for every user 5 times, and each time user n produces 5 genuine signatures and 5 forgeries for user $n - i$, where $i = 1..2$. As a result every user is providing 25 genuine signatures (5 × 5) and 25 forgeries (5 per user $n - 1$, 5 per user $n - 2$, ..., 5 per user $n - 5$). In this scheme, every user between writing a set of 5 genuine signatures is forced to change his style in order to produce forgeries. This increases variability between each set of genuine signatures in relation to variance inside each set. The total amount of signatures available in MCYT-100 is 5000 ($100 \times 25$ genuine + $100 \times 25$ forgery).

## 2.5.4 BIOMET

BIOMET is multimodal database that contains audio, face images, hand images, fingerprints, and signatures [13]. In contains online signatures registered with WACOM Intuos2 A6 tablet. Signature characteristics were captured with 100 Hz sampling rate and include $X$-coordinate, $Y$-coordinate, pen altitude, azimuth, and pressure. The signatures were captured in two sessions with the interval of 5 months. During the first session 5 genuine and 6 forgeries were registered for each individual. The forgers had access to the image of a genuine signature that they tried to imitate. In the next session additional 10 genuine samples and 6 forgeries were acquired per person. The database contains complete sets of samples for 84 individuals.

## 2.6 Commercial Software

On the market there are several systems that facilitate signing of documents and signature verification. The functionality of commercial packages include

- Location and extraction of offline signature from scanned documents.
- Ability to capture signature using typical input devices for online registration.
- Embedding captured signature into electronic documents in portable formats (such as *.pdf—portable document format) with an option to lock documents for changes after signing.
- Verification of captured or embedded signatures against reference samples.
- Integration with digital certificates for encryption and protection against signed document changes.

## 2.6.1 SOFTPRO

A SOFTPRO company (http://www.softpro.de/ [accessed on January 14, 2014]) offers a suite of products that allow for signing documents, signature verification, and integration with document management systems. Among institutions that use SOFTPRO solutions that facilitate signing documents and signature verification are insurance agencies, banks, and logistics companies.

Embedding a signature in a document with biometric (online) data for its verification requires additional protection. The scheme of protecting a signature together with signed document that is used in SOFTPRO products involves usage of public and private keys of Public Key Cryptography Standards as defined by RSA Security. In order to protect data two pairs of keys are required:

1. A pair used for biometric data encryption, private key of this pair should be stored in a secured location, public key is configured to be used by the system as public trusted third party key.
2. A pair used for protecting the integrity of document, can be self-signed and deleted afterward.

The following steps are used to protect the signature data and signed document:

1. Transmission of online signature data from acquisition device over encrypted channel: AES-128 or device embedded encryption method is used.
2. Storing signature image computed from online data in the document being signed.
3. Storing biometric data in the document. Before storing, biometric data is encrypted using RSA up to 4096 and AES-256 CBC (the public key from key pair 1 is used for signature encryption—public trusted third party key).
4. Calculation of SHA-256 hash for the document to guarantee the integrity of document with encrypted biometric data. The hash key is signed with private key of key pair 2.
5. The hash value and relevant public key of key pair 2 are stored in the document. To bind biometric data to the document another hash is calculated utilizing SHA-256 on top of the previous hash and biometric data.

### 2.6.2 ParaScript

ParaScript (http://www.parascript.com/ [accessed on January 20, 2014]) offers signature verification software for offline and online data. Offline signature verification is based on features such as comparison of geometric shapes, fragments, and trajectories. When online data is acquired, pressure, speed, and tension are analyzed. For static documents additional software is provided that allows to locate and extract signature in typical forms and bank checks.

### 2.6.3 SQN Banking Systems

SQN Banking Systems (http://www.sqnbankingsystems.com [accessed on January 20, 2014]) provides solutions for bank institutions for signature processing and verification. Their products related to signature verification allow for extracting applicable signatures, comparing the current signature to those in a verified database, assigning a confidence score enabling suspects, and reference signatures to be displayed in a workflow application.

## 2.7 A Review to Signature Recognizers

The architecture of signature recognition system does not differ much from a typical biometric system. Its main stages are data acquisition, preprocessing, feature extraction and classification. At the data acquisition stage, the signature data is usually obtained by means of a scanner, a tablet, a specialized signature pad, or a digital camera. After digitalization, additional preprocessing is required to prepare data for the next stage—feature extraction.

The feature extraction methods are responsible for constructing the feature vector—a mathematical representation of the signature instance that contains significant information required for proper classification. The classification process is used for one of the two tasks: identification and verification. The aim of verification is to decide whether the given signature is genuine instance belonging to a particular user or if it does not match his/her reference samples and should be treated as a forgery. In this mode of operation the system compares tested signature with a model of a particular user (1 to 1 comparison).

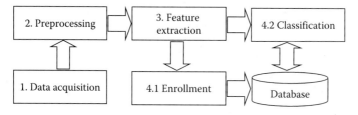

**Figure 2.4**  Biometric signature system architecture.

During the identification the system finds individual whose signature best matches given sample. In this case, the tested signature is compared with all user's models (1 to $N$ comparisons, where $N$ is number of users enrolled in the system).

Another mode of systems operation is enrollment—during this process the genuine signatures of a particular user are registered as his reference samples (Figure 2.4).

The following sections briefly describe the commonly used approaches at different stages of signature biometric systems.

### 2.7.1  Data Acquisition

During data acquisition handwritten signatures are collected for further processing. There are two ways in which data can be acquired: offline—where the input of the system are static images of handwritten signatures; online—registers the act of signing that includes both the image and dynamics of writing. Due to specific nature of data, offline and online signatures usually require different methods at each stage of the biometric system.

For offline processing static images of handwritten signatures can be obtained by means of scanning devices from original documents. In order to extract characteristic features of signature the acquired image should have proper quality. The resolution should be not less than 300 DPI and images must be stored in a format without lossy compression.

In case of online signature, the acquisition must be performed during signing in order to register the way the signature was given—dynamics of writing the signature. In most such acquisition is being made by making use of devices with touch-sensitive surface which can register pen movements. Nowadays, there is a wide range of electronic devices equipped with touch-sensitive surfaces such as graphical tablets, mobile tablets, smartphones, and so on. In addition, one can find specialized tablets (often called signature pads) designed for capturing online signatures.

An example of a graphical tablet that is used in many investigations is WACOM Intuos series. Intuos Pro model allows for the registration of the following parameters (with 200 Hz sampling rate):

- $x$-Coordinate, $y$-coordinate (5080 lines per inch)
- Pressure (2048 levels)
- Azimuth, altitude

The main drawback of using this kind of device is that the signer has to look at separate screen in order to see his writings. This may lead to different characteristics of signatures registered in this way from signatures given on a plain paper.

One solution to this problem is to apply an inking pen that can be used to simultaneously write on a paper and allow for online registration by a tablet placed beneath a paper form. Other option is to use a special pen that is capable of registering dynamics of writing without using a touch-sensitive surface. An example such technology is WACOM Inking pen that allows for registration of $X$-coordinate, $Y$-coordinate, and pressure while writing on a paper. Such devices can also be used in applications where a paper document is required (e.g., a receipt should be given out to the client or an agreement should be signed on paper document due to legal reasons). Some models of pens can also register a force inflicted on the pen ball in three dimensions: two orthogonal force components in the plane of a paper ($x$- and $y$-force) and one component along the pen direction (representing downward pressure when pen is in upright position). In work [14] force components where used for online signature verification.

Apart from general-purpose tablets, there are specialized devices for signature acquisition (Figure 2.5) such as Wacom STU-500 and ePadLinkePad®. In case of STU-500 model, the writing surface is touch-sensitive LCD panel that makes the process of signing more natural to the writer because the signature line is displayed below pen tip which imitates writing on a paper. The STU-500 records the following parameters (with 200 Hz sampling rate):

- $x$-Coordinate, $y$-coordinate (2540 lines per inch)
- Pressure (512 levels)

(a)                              (b)                              (c)

**Figure 2.5** Examples of digital tablets for online signature registration: (a) WACOM Intuos Pro graphical tablet, (b) WACOM STU-500 LCD signature tablet, and (c) ePadLink signature tablet.

Signature pads with LCD screen can be used in applications when there is no need to store documentation in a paper form. LCD can display document that requires a signature and after signing both document and signature can be stored together in a single digital file.

Nowadays, popular mobile devices such as mobile tablets and smartphones are also equipped with touch-sensitive screens that allow for online signature registration. However, the most of them provides only $X$- and $Y$-coordinates (with some exceptions such as Galaxy Note tablets and smartphones that can also register up to 1024 levels of pressure) with lower resolution and sampling rate than graphic or specialized tablets.

Online signature can be also registered as a video sequence using digital camera [15]. In this case from tracing a pen tip dynamics of trajectory can be acquired in a form of $X$ and $Y$ coordinate samples. The sampling rate is constrained by the number of frames per second in a video sequence, the resolution of $x$ and $y$ values depend on the resolution of picture and distance of camera from the writing surface.

Registration of online signature can be also integrated into interactive kiosks stations that can be used in public places (Figure 2.6). Such devices often include credit card readers and can be used as POS terminals.

**Figure 2.6** Examples of Kiosk terminals with capability of online signature registration.

Data recorded for online signature takes the form of series of values, one series for each characteristics such as $X$, $Y$ coordinates or pressure. Such information can be stored in arbitrary format depending on requirements of particular application. In order to facilitate exchange of data between different biometric systems international standards can be used.

ISO/IEC 19794-7:2007 specifies two data formats for signature data captured in the form of time series using devices such as digitizing tablets or advanced pen systems: format for general use and its compacted version for use with smart cards or other tokens. Both data formats can be used for raw data acquired directly from input device and for time-series features that are used at classification stage.

When capturing online signature it is crucial to protect against exposure of biometric data to potential forgers. The protection of data should start with acquisition device and transmission of data from the device to its destination (usually PC). The specialized signature pads such as WACOM STU-530 have embedded AES 256/RSA 2048 encryption that is used during transmission of data from acquisition device.

### 2.7.2 Preprocessing

The aim of preprocessing is the transformation of raw data obtained from acquisition device into a form that is more adequate for feature extraction. The basic tasks performed at this stage include localization of signature in a document, extraction of signature line from background, transformation of the data, and normalization.

The localization and segmentation of offline signatures from acquired scans can be easily implemented if certain constraints can be applied on the position of the signature inside the analyzed document. In many forms signature is given at fixed position that could be included as a configuration parameter.

Extraction of signature image from more complex documents and their backgrounds (e.g., bank check in Figure 2.7) is not trivial and one can find several works devoted to this subject. Typical approaches are subtraction of a blank form image from filled document [16], extraction of handwriting based on its characteristics [17], using general image segmentation methods such as connected components

**Figure 2.7** Eample of a scanned signature with nonuniform background.

labeling [18], or region growing [19]. Another problem is the noise and defects caused by poor quality of documents and the scanning process. The quality of images can be improved using standard filtering techniques (e.g., median filter).

In the case of dynamic signatures registered by means of PC tablet or specialized signature pad, the preparation of input data for feature extraction process is much simpler. The data provided by these devices can be directly transmitted to the feature extraction step—there is no need for signature localization and segmentation from the background. Various parameters such as trajectory or pressure are recorded with high precision and there are no artifacts that can damage the signature line.

For online signatures several works use interpolation to resample acquired data using fixed time interval or fixed distance along signature line (Figure 2.8). Some researchers suggest, however, that such a process can lead to loss of important data for verification, for example, characteristics of pen speed and moments of rapid changes in direction can be distorted.

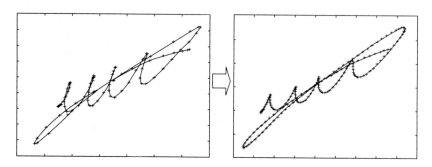

**Figure 2.8** Interpolation of online signature trajectory. (Data from M. Martinez-Diaz et al., On the effects of sampling rate and interpolation in HMM-based dynamic signature verification, *9th International Conference on Document Analysis and Recognition*, IEEE, Paraná, Brazil, pp. 1113–1117, 2007. With permission.)

The recorded signatures may be scaled, shifted, and rotated with respect to the origin. These factors may affect the recognition process, because they introduce additional variability among signatures belonging to the same person. For this reason, a frequently used component of the preliminary data processing is the process of normalization. There are many methods of normalization, the most common are linear scaling [21], shifting in respect to the center of mass, and the elimination of the trend line using linear regression [22]. In addition to using an explicit normalization, representations of features that are invariant with respect to affine transformations are used [18].

When choosing a normalization algorithm or making decision about the use of invariant features, it is important to not to lose the signature characteristics that can be used during recognition process. For example, in case of scaling it is important to preserve signature aspect ratio. Decision to use particular normalization technique should be verified by evaluating system error. Moreover, applicability of certain techniques may depend on the source of acquired signatures. For instance, if all signatures come from a form with fixed window for placing a signature, the size of the signature shape could be an important feature.

Static signature image, after separation from the background of the document and the removal of noise, is still a complex graphic object composed of thousands of points. To facilitate the analysis of such a complex object, the thinning technique can be applied. The aim of thinning is to obtain a 1-pixel wide line that preserves the complete topology of the original shape (see Figure 2.9).

Thinning signature line facilitates the work of algorithms that track the direction and shape of individual segments making up its structure. Some methods use this representation in an attempt to reconstruct the writing trajectory—the sequence of movements of the pen tip when plotting the signature (Figure 2.10). Information about trajectory is important in the task of verification, because it allows to create a more accurate writer's model [1]. The signature line presented as a one-dimensional sequence of values would allow for an analysis of the order in which individual features occur in time—an analysis of signature dynamics on the basis of the static

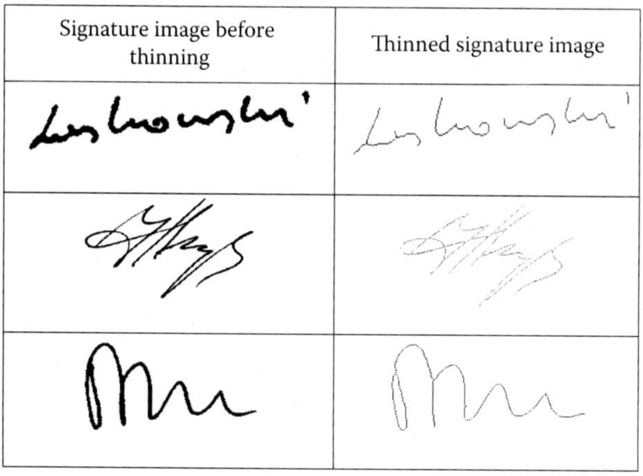

| Signature image before thinning | Thinned signature image |
|---|---|
| *Lushoushi'* | *Lushoushi'* |
| | |
| | |

**Figure 2.9** Examples of signatures before and after thinning.

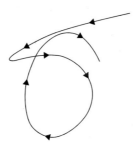

**Figure 2.10** An example of trajectory reconstruction. (Data from E. M. Nel et al., Estimating the pen trajectories of static signatures using Hidden Markov Models, *IEEE Transactions on Pattern Analysis and Machine Intelligence*, 27, 1733–1746, 2005. With permission.)

signature image. Research related to this subject can be divided into two categories:

- Reconstruction of offline signature trajectory using information that were obtained only from static images
- Reconstruction of offline signature trajectory based on data from previous online acquisitions

Lee and Pan presented one of the first algorithms for reconstruction of the trajectory. In their method the base track was chosen heuristically with a set of rules [24], which determined the order of points on thinned signature line. These rules were developed by analyzing the way in which a person given static examples of handwriting tries to reconstruct its trajectory.

Another approach is presented in the work of Lau Yuen and Tang [25]. Tracking process is based on a developed model of signing and a process of optimization that searches the best trajectory that could be produced be the assumed model.

Regardless of the method that is used, the trajectory of signing cannot be reliably reconstructed in all cases using information only from static images. In order to improve reconstruction process, examples of genuine writing sequences must be provided. In works [23, 26] the authors utilize dynamic examples registered previously using tablet device as guidelines during trajectory reconstruction for newly acquired offline signatures.

One of the important issues faced by the signature recognition systems is relatively large variation between the authentic signatures of the given person. This translates to the difficulty in determining the stable elements or stable shape description which could be referenced during verification task. In order to reduce the negative effects of this variation, in many systems, before the step of feature extraction, a segmentation of signature line is performed. The aim of this process is division of the line into segments that reflect relevant and stationary components of the signature. One of the simplest techniques for signature segmentation is dividing its image based on connectivity of signature line (Figure 2.11). This technique was used in work [27].

Brault and Plamondon in [28] developed a two-step algorithm for signature line division based on points of so-called highest perceptual importance. At the first stage, for each point of signature line a value of a function is assigned that weights its importance based on its neighboring points along signature line. During the next stage local maxima are identified and are used as points of division into segments. The segments obtained by this method have uniform curvature, while the points of intersection usually represent changes in direction (Figure 2.12).

In work [29] authors extracted segments from outer and inner contours of signature image. Every contour was traced in clockwise direction starting from the leftmost pixel. The contour line was divided into

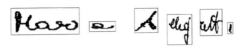

**Figure 2.11** Segmentation into connected components.

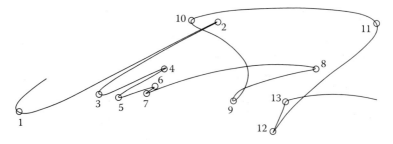

**Figure 2.12** An example of signature line segmentation. (Data from J. J. Brault and R. Plamondon, Segmenting handwritten signatures at their perceptually important points, *IEEE Transactions on Pattern Analysis and Machine Intelligence*, 15, 953–957, 1993. With permission.)

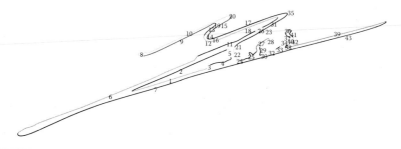

**Figure 2.13** Segmentation of signature outer and inner contours. (Data from Y. Rekik et al., A comparison of feature extraction approaches for offline signature verification, *International Conference on Multimedia Computing and Systems*, IEEE, Ouarzazate, Morocco, pp. 1–6, 2011. With permission.)

strokes at points which corresponded to changes of direction along the *X*-axis or *Y*-axis or to high changes of slant directions (Figure 2.13).

Another method was used in work [30]. The signature was divided into rectangular blocks with the size of $50 \times 50$ pixels, forming a grid that was overlaid onto image (Figure 2.14). The local features describing a signature were computed separately for each of these boxes.

In case of online signature, the segmentation process can take advantage of dynamic features captured during act of signing. Most of tablets register moments of touching the tablet surface (pen down event) and lifting the pen (pen up event) during writing. Such moments may be used to divide signature into components that were written separately (Figure 2.15).

Another segmentation scheme used in works [31,32] utilizes the characteristics of pen tip velocity computed from data registered during act of signing. In this approach the signature line is divided into so-called strokes, that are formed from parts of the trajectory bounded by two consecutive local velocity minima. As shown in Figure 2.16,

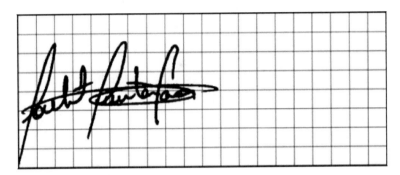

**Figure 2.14** An example a grid-based segmentation. (From C. Santos et al., An off-line signature verification method based on the questioned document expert's approach and a neural network classifier, *Proceedings of the 9th International Workshop on Frontiers in Handwriting Recognition*, IEEE, Tokyo, Japan, pp. 498–502, 2004. With permission.)

**Figure 2.15** Online handwritten signature segmented with pen-down (dot) and pen-up (star) events. (From D. Impedovo and G. Pirlo, Automatic signature verification: The state of the art, *IEEE Transactions on Systems, Man and Cybernetics, Part C: Applications and Reviews*, 38, 609–635, 2008. With permission.)

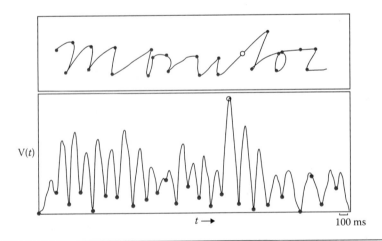

**Figure 2.16** An example of velocity-based segmentation: (a) segmented trajectory and (b) velocity characteristics with local velocity minima. (From L. Schomaker, From handwriting analysis to pen-computer applications, *Electronics & Communication Engineering Journal*, 10, 93–102, 1998. With permission.)

places where minima occur correspond to parts of signature line with highest curvature and changes of direction.

The segmentation process executed on different instances of signatures given by the same user often produces different segments for each instance. Due to this fact it may be difficult to find corresponding segments in signatures being compared. In work [33] a method based on DTW measure was proposed, in that given segmentation points determined for one signature finds the corresponding segments in the second signature that creates best alignment between the two. Similar process were used in work [34], but with the segmentation driven by reference model (model-guided segmentation).

## 2.7.3 Feature Extraction

The data obtained from acquisition and preprocessing needs to be prepared for classification process. In most cases it means extraction of information that best describes particular signature and discarding any superfluous data that can be safely ignored. While there are many different approaches to the selection of characteristic features of handwritten signature, most of the features can be divided into two categories: functional and parametric.

Functional features can be represented as a function $f(t)$ of specific argument (usually time for online signature or coordinate for offline). Basic characteristics acquired for during online registration can be described as functional features:

- $X$-Coordinate

$$X = (x_1, x_2, ..., x_N), \quad x_i = x(t_i) \tag{2.1}$$

- $Y$-Coordinate

$$Y = (y_1, y_2, ..., y_N), \quad y_i = y(t_i) \tag{2.2}$$

- Pressure

$$P = (p_1, p_2, ..., p_N), \quad p_i = p(t_i) \tag{2.3}$$

- Elevation

$$L = (l_1, l_2, ..., l_N), \quad l_i = l(t_i) \tag{2.4}$$

- Azimuth

$$Z = (z_1, z_2, ..., z_N), \quad z_i = z(t_i) \tag{2.5}$$

Using above features other characteristics can be computed such as velocity, acceleration, and direction:

- Velocity along $X$-axis

$$V^x = (v_1^x, v_2^x, ..., v_N^x), \quad v_i^x = \frac{x(t_{i+1}) - x(t_{i-1})}{t_{i+1} - t_{i-1}} \tag{2.6}$$

- Velocity along $Y$-axis

$$V^y = (v_1^y, v_2^y, ..., v_N^y), \quad v_i^y = \frac{y(t_{i+1}) - y(t_{i-1})}{t_{i+1} - t_{i-1}} \tag{2.7}$$

- Total velocity

$$V = (v_1, v_2, ..., v_N), \quad v_i = \sqrt{(v_i^x)^2 + (v_i^y)^2} \tag{2.8}$$

- Acceleration along $x$-axis

$$A^x = (a_1^x, a_2^x, ..., a_N^x), \quad a_i^x = \frac{v_{i+1}^x - v_{i-1}^x}{t_{i+1} - t_{i-1}} \tag{2.9}$$

- Acceleration along $y$-axis

$$A^y = (a_1^y, a_2^y, ..., a_N^y), \quad a_i^y = \frac{v_{i+1}^x - v_{i-1}^x}{t_{i+1} - t_{i-1}} \tag{2.10}$$

- Total acceleration

$$A = (a_1, a_2, ..., a_N), \quad a_i = \sqrt{(a_i^x)^2 + (a_i^y)^2} \tag{2.11}$$

- Tangent of direction

$$\theta = (\theta_1, \theta_2, ....\theta_N), \quad \theta_i = \frac{y(t_{i+1}) - y(t_{i-1})}{x(t_{i+1}) - x(t_{i-1})} \tag{2.12}$$

The dimension of functional features ($N$) depends on the sampling rate and the time of signing. Parametric features are computed as various statistics that describe features of a signature. In case of online signatures such features may include [35,36]:

- Total signing duration
- Total pen down duration
- Number of pen up events
- Number of pen down events
- Time of second pen down event
- Number of registered samples
- Average of velocity $v$

- Maximum of $v$
- Time when maximum of $v$ occurred
- Number of sign changes of horizontal velocity $v^x$
- Number of sign changes of vertical velocity $v^y$
- Standard deviation of $v^x$
- Standard deviation of $v^y$
- Duration of $v^x > 0$
- Duration of $v^y > 0$
- Duration of $v^x < 0$
- Duration of $v^y < 0$
- Average positive $v^x$
- Average positive $v^y$
- Average negative $v^x$
- Average negative $v^y$
- Number of $v^x = 0$
- Number of $v^y = 0$
- Maximum of $v^x$ − minimum of $v^x$
- Maximum of $v^y$ − minimum of $v^x$
- Standard deviation of horizontal $a^x$
- Standard deviation of vertical $a^y$
- Initial direction $= \theta_0$
- Direction from first pen down to last pen up
- Direction from first pen down to second pen down
- Direction from first pen down to second pen up
- Maximum of pressure $p$
- Average of $p$
- Maximum of azimuth $z$
- Minimum of $z$
- Maximum of altitude $l$
- Average of $l$
- Maximum of $x$ − minimum of $x$ (W)
- Maximum of $y$ − minimum of $y$ (H)
- Standard deviation of $x$
- Standard deviation of $y$
- Number of local maxima of $x$
- Number of local maxima of $y$
- Aspect ratio $W/H$

Offline signatures features are based on information obtained from signature image. There are various methods for feature extraction and examples of popular approaches are given in the following pages. Offline signature features are usually categorized as global or local. Global features are computed using whole signature image and represent characteristics of its entire shape. They allow to distinguish signatures of different individuals that vary significantly in their appearance and capture holistic features that should be stable and therefore present in all genuine signatures given by the same person. Local features are determined for specific parts of the signature, usually resulting from segmenting the signature line or image. As a result, they contain detailed information on the shape and structure of particular components and are important for forgery detection.

*2.7.3.1 Graphology Based* In work [6] the authors presented a set of features derived from concepts present in graphology and graphometry. The following characteristics of handwriting as defined in graphology and graphometry where analyzed for their applicability in automated systems: order—distribution of graphical elements, proportion—symmetry of writing, dimension—aspect ratio of letter height and width, pressure—thickness of signature related to the pressure of pen, constancy—describing the speed and intensity of writing, form—the type strokes that are dominant in writings (e.g., horizontal, vertical, and rounded), characteristic gestures—elements that the writer repeats periodically in a specific manner (e.g., how he starts/finishes writings), occupation of the space—describes the way a particular writer uses a space available for writing.

Based on above characteristics a set of features types that could be extracted for signatures using automated methods was presented [6]:

- *Caliber*: Determines the height and width of the signature.
- *Proportion*: Describes geometric regularity of a signature.
- *Spacing*: Contains information on empty regions (spaces) between signature components.
- *Alignment to baseline*: Depicts position of signature relative to the baseline.
- *Progression*: Describes characteristics of writing related to its speed.
- *Form*: Conveys the characteristics of strokes in a signature image.
- *Slant*: Describes the slant of signature.

The caliber, propagation, spacing, and alignment features were categorized as static features as they describe the occupation of the graphical space, while progression, form, and slant were called as pseudodynamic due to their dependence on dynamics of signing.

In order to extract above characteristics a signature image was segmented using grid rectangular cells (with the proposed size of 16 × 40 cells). For each cell separate a set of pseudodynamic features where computed and included in signature feature vector. Information on static features (caliber, proportion, spacing, and alignment to baseline) was incorporated implicitly through the use of pseudodynamic features computed for each segment of grid.

In order to represent progression three features where calculated: density—number of pixels forming a signature line in each cell, the distribution of pixels—each cell was divided into 4 for zones and the width of the stroke in a cell was computed in four directions constrained to the created zones (Figure 2.17), and the progression—calculated based on a number of times that longest stroke in a cell changed its direction.

The signature slant was computed using a two-step process: calculation of global slant for entire image and local slant for each cell. In order to describe the form feature for each cell a concavity-based measure was used.

*2.7.3.2 Shape Context Based*   In works [5,37] a description of offline signature based on shape contexts was proposed. The shape context

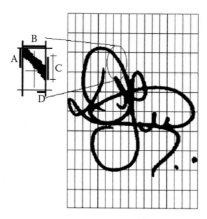

**Figure 2.17**   An example of computing distribution of pixels. (From L. S. Olivera et al., The graphology applied to signature verification, *12th Conference of the International Graphonomics Society*, pp. 286–290, 2005. With permission.)

algorithm [38] allows for measuring shape similarity by solving the so-called correspondence problem between two objects ($R$, $S$) and finding an aligning transform. Each of the two signatures, whose shapes are compared, is represented by a set of points:

$$R = \{r_1, r_2, ..., r_N\}$$

$$S = \{s_1, s_2, ..., s_N\},$$

(2.13)

For each point on the reference signature ($R$), a corresponding point on the second signature ($S$) is found. In order to find the corresponding pairs each point is described by a shape context descriptor. The shape context descriptor contains information about the configuration of the entire shape relative to the point being described (Figure 2.18). It is formed by computing a coarse histogram representing distribution of points comprising the signature line relative to the reference point. Bins used in the histogram calculation are uniform in log-polar space:

$$h_k(s_i) = \# \{s_j \in S : s_j \in bin(k), j = 1...N \wedge j \neq i\} \quad (2.14)$$

where $h_k(s_i)$ is the $k$th bin of log-polar space histogram describing distribution of points in $R$ relative to point $s_i$.

Descriptor possesses much more information on the role of a point in the object than its position given by the coordinate system. It allows for finding better correspondence between points of compared objects, especially in the case of offline signatures where the points extracted from an image do not form ordered sequence due to lack of information on the writing trajectory.

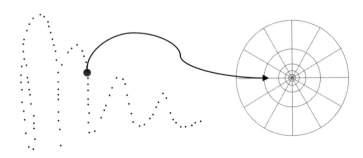

**Figure 2.18**  Description of signature point using log-polar space histogram. An example of computing distribution of pixels. (From M. Adamski et al., Signature system based on extended shape context descriptors, *International Conference on Biometrics and Kansei Engineering*, IEEE, Tokyo, Japan, pp. 267–272, 2013. With permission.)

As a result the signature is described by a set of histograms:

$$S = \{h(s_1), h(s_2), ..., h(s_N)\} \tag{2.15}$$

where $h(s_i)$ is the histogram describing distribution of signature points relative to point $s_i$.

*2.7.3.3 Contour Based*   In work [39] a feature vector was constructed using signature contour. Contour tracing algorithm follows the boundaries comprising object image and collects coordinates of their consecutive points. After their extraction, the signature contours where segmented into upper and down segments. Two pixels: one with the minimum (left-most) and one with the maximum (right-most) value of $x$-coordinate were selected as the points of division. Boundary pixels starting from the left-most to the right-most pixel counting in clockwise direction formed the upper segment, pixels from left-most to the rightmost pixel in counter clockwise direction were included in the down segment. If a signature consisted of several separated components, corresponding segments of each component were concatenated in the order determined by the $x$-coordinate value of their left-most pixel. The feature vector was created from $y$-coordinates of boundary pixels sampled along upper and then bottom concatenated segments. In addition the number of internal boundaries (equal to number of closed loops) was included in the feature vector (Figure 2.19).

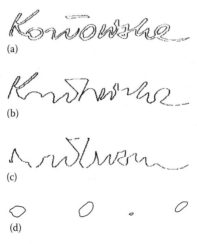

(a)

(b)

(c)

(d)

**Figure 2.19**   Examples of full contour (a), upper concatenated contour (b), bottom concatenated contour (c), and internal contours (d).

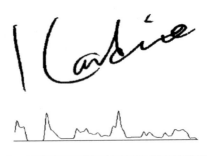

**Figure 2.20**    A signature (a) and its vertical projection (b). (From B. Fang et al., Off-line signature verification by the tracking of feature and stroke positions, *Pattern Recognition*, 36, 91–101, 2003. With permission.)

(a)                              (b)                              (c)

**Figure 2.21**    An example of a signature (a), its thinned version (b), and approximation with straight lines (c). (From B. Fang et al., Off-line signature verification by the tracking of feature and stroke positions, *Pattern Recognition*, 36, 91–101, 2003. With permission.)

*2.7.3.4 Projection Based*    The signature image can be also represented using vertical and horizontal projections. The vertical projection is formed by counting pixels of the signature line in every column of the image (Figure 2.20). The horizontal projection is computed by counting pixels in the image rows. Such projections are one-dimensional sequences that can be incorporated directly in the feature vector. Example of this approach can be found in work [40].

*2.7.3.5 Curvature Based*    Another method of signature description proposed in [40] was based on approximation of signature curve with short straight lines (Figure 2.21). The signature image was characterized by a set of lines, where each line is characterized by its slope and position vector of its midpoint. To determine line approximations the signature image was initially thinned.

*2.7.3.6 Radon Transform Based*    In work [9] the feature vector was constructed by applying discrete radon transform to signature image. In order to obtain the feature vector the signature image was projected using $\theta$ angles with $\beta$ projection beams per angle.

$$R_j^i = \sum_{k=1}^{K} w_{ijk} * p_k \qquad (2.16)$$

where:

$i = 1...\Theta, j = 1...\beta$

$p_k$—intensity of $k$th pixel in signature image

$w_{ijk}$—contribution of $k$th pixel into $j$th bean for angle $i$

After obtaining vectors $R^i$ the dimensionality of each $R^i$ was changed from $\beta$ into $d$ by decimating zero values and extending or shrinking resulted vectors into dimension of $d$ by using interpolation technique. After additional normalization the values of $R^i$ were incorporated into feature vector.

Figure 2.22 shows signature image with two projections at $0°$ and $90°$ degrees (a) and DRT of that image presented in gray scale (b).

*2.7.3.7 Hough Transform Based* The authors of work [41] assumed that the signature consisted of many straight lines which are unique for each person. To create a feature vector containing description of these lines the Hough transform was used. At first step the signature images where normalized to the size of $128 \times 128$ pixels and binarized using thresholding technique. Next, the signature objects where translated to the left and bottom image boundary and thinned using morphological erosion. At this stage the Hough transform was applied with the accumulator of 4096 elements. The feature vector was formed by using the number of votes directly from the accumulator (Figure 2.23).

*2.7.3.8 Texture Based* In [42] the textural features extracted with local directional pattern (LDP operator were used to describe offline signature image). LDP operator describes image pixel in its

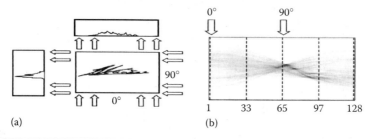

(a)      (b)

**Figure 2.22** Signature image with two projections at $0°$ and $90°$ (a) and DRT of that image presented in gray scale (b). (From J. Coetzer et al., Offline signature verification using the discrete radon transform and a Hidden Markov Model, *EURASIP Journal on Applied Signal Processing*, 2004, 559–571, 2004. With permission.)

| Hough space ($\rho,\theta$) | | |
|---|---|---|
| No. of feature | $\rho$ (rho) | $\theta$ (theta) | # of votes |
| 1 | 1 | 1 | 32 |
| 2 | 1 | 2 | 79 |
| 3 | 1 | 3 | 64 |
| . | . | . | . |
| . | . | . | . |
| 256 | 4 | 1 | 72 |
| . | . | . | . |
| 4096 | 64 | 64 | 18 |

**Figure 2.23** Feature vector based on the Hough transform. (Data from T. Kaewkongka et al., Off-line signature recognition using parameterized Hough Transform, *Proceedings of the Fifth International Symposium on Signal Processing and Its Applications, ISSPA '99.*, IEEE, Brisbane, Australia, pp. 451–454, 1999. With permission.)

$$\begin{bmatrix} -3 & -3 & 5 \\ -3 & 0 & 5 \\ -3 & -3 & 5 \end{bmatrix} \quad \begin{bmatrix} -3 & 5 & 5 \\ -3 & 0 & 5 \\ -3 & -3 & -3 \end{bmatrix} \quad \begin{bmatrix} 5 & 5 & 5 \\ -3 & 0 & -3 \\ -3 & -3 & -3 \end{bmatrix} \quad \begin{bmatrix} 5 & 5 & -3 \\ 5 & 0 & -3 \\ -3 & -3 & -3 \end{bmatrix}$$

East $M_0$     Northeast $M_1$     North $M_2$     Northwest $M_3$

$$\begin{bmatrix} 5 & -3 & -3 \\ 5 & 0 & -3 \\ 5 & -3 & -3 \end{bmatrix} \quad \begin{bmatrix} -3 & -3 & -3 \\ 5 & 0 & -3 \\ 5 & 5 & -3 \end{bmatrix} \quad \begin{bmatrix} -3 & -3 & -3 \\ -3 & 0 & -3 \\ 5 & 5 & 5 \end{bmatrix} \quad \begin{bmatrix} -3 & -3 & -3 \\ -3 & 0 & 5 \\ -3 & 5 & 5 \end{bmatrix}$$

West $M_4$     Southwest     South $M_6$     Southeast $M_7$

**Figure 2.24** Kirsch filter masks used to compute LDP. (Data from M. A. Ferrer et al., Signature verification using local directional pattern [LDP], *IEEE International Carnahan Conference on Security Technology,* IEEE, San Jose, CA, pp. 336–340, 2010. With permission.)

8-neighborhood by responses of 8 Kirsch filter masks (Figure 2.24) applied at the pixel location.

The three highest response values that denote dominant directions and are encoded by 1, the rest is set to zero. As a result an 8-bit code is created for every image pixel. To reduce the dimensionality of data (the number of codes is equal to number of pixels) the image was divided into 12 blocks, and for each block a histogram of LDP values was calculated. At the next step, the histograms were merged into single sequence and the discrete cosine transform (DCT) was applied to it. The feature vector was formed by selecting the first 168 components of DCT transform.

*2.7.3.9 Wavelet Transform Based* The feature vector describing the signature line can be also obtained based on coefficients of the wavelet transform. Such representation allows to include signature features at different resolutions that capture both global shape characteristics and local details. In work [43] the wavelet transform was applied to the entire signature image. In other approaches one-dimensional representations such as signature projections were used as an input to the wavelet transform [44].

### 2.7.4 Classification

In the classification stage the system issues a decision based on the feature vector of examined signature and reference data registered during enrollment. Methods that are used in this stage can be divided into two categories: template matching and statistical classification.

#### 2.7.4.1 Template Matching

In template matching the classification process is based on assessment of similarity between reference and verified signatures. This assessment is usually carried out by means of a specific function that evaluates similarity or produces a measure of distance between compared signatures. Template matching approach can be used when a single or several reference signatures per user are available. In case of a single reference signature, the direct comparison of the feature vectors obtained for the reference and verified signatures produces a value that can be used to issue the final decision.

When several reference signatures per user are available, the template matching can be utilized as a distance measure in the $k$-nearest neighbors framework. Another approach is to select the most representative reference or construct average model for one-to-one comparison.

Template matching methods are mostly used with functional features—the basic representation of online signature characteristics. Such features describe signature at local level and have highest level of detail. Due to the fact that such representation has high dimension, the classification based on template matching techniques can be computationally intensive. Relatively high complexity of the matching algorithms only adds to this negative effect and may require additional techniques such as initial pruning when many comparisons are needed (e.g., in identification task).

##### 2.7.4.1.1 DTW

Dynamic time warping (DTW) is the most frequently used template matching technique for online signature verification. It allows to align functional features of compared signatures by nonlinear time-axis transformation. In many studies this approach gave very good results that often exceeded results obtained for systems that used statistical classification with parametric features.

In basic application, DTW can be directly utilized as a distance measure that compares functional features of the test and reference signatures. The distance is computed on sequences initially aligned by the DTW algorithm.

Other approaches make use of the alignment produced by DTW as a basis for more advanced analysis. In work [45] the linear regression was computed on aligned sequences in order to obtain similarity measure for online signature verification. In [40] the DTW was used for static signature profiles. In this case the authors proposed a distance that was based on local displacements between profiles of tested and reference signatures computed by DTW alignment technique. As the authors of [40] explain, the statistics of displacement corresponds to positional variations of the signature strokes and are useful for making decision whether a given signature is authentic or not.

*2.7.4.1.2 Other Techniques*  In work [46] a similarity measure based on logarithmic spectrum was used for online signature verification. The logarithmic function was applied to the frequency spectrum obtained by fast Fourier transform from functional data. Then, the principal component analysis was performed to construct a feature vector from logarithmic spectrum coefficients. The value of similarity measure between feature vectors of reference and verified signatures was compared with the threshold to determine the signature authenticity.

The method proposed in [47] was based on assumption that each writer has his own statistics of handwriting displacement and therefore a statistical displacement analysis can be used for writer verification. The distance measure between two signatures was based on normalized displacement function and statistics of displacement.

In work [48] the relaxation matching was used for offline signature verification. The signatures were represented with directional features in a form of DF curves obtained by contour tracing. Each individual curve in the test signature was accepted or rejected based on comparison with corresponding curves in reference model. A global distance value was calculated based on matched curves.

*2.7.4.2 Statistical Classification* In literature of automatic handwritten signature recognition one can find applications of almost all commonly used statistical classifiers. The most popular are neural networks, hidden Markov models (HMMs), support vector machines (SVMs), and Bayesian networks. Statistical classifiers are used with parametric features. Both global and local features are utilized to construct feature vectors. In order to train statistical classifiers relatively large database of reference samples is needed. At least five reference signatures for each user is required in order to conduct learning. This narrows applications of this approach. On the other hand, statistical classifiers offer better performance and competitive success rate as compared with template matching techniques.

*2.7.4.2.1 Neural Networks* Application of neural networks to offline signature classification can be found in [30], where single multilayered neural network (MLP) was used in a role of a document expert. In this case the classifier was given five reference samples of a particular individual together with a test signature. Provided with this input, the classifier issued a decision whether the test sample matched the reference samples (successful verification) or not (rejection). The neural network used multilayer perceptron (MLP) topology with 640 neurons in the input layer, 4–16 neurons in the hidden layer and 2 neurons as the output.

In study [49], MLP network was used for dynamic signature identification and verification. In the network that was responsible for identification the number of neurons in the output layer corresponded to the number of persons who provided authentic signatures. In the experiment the database contained signatures of 16 people, as the result the output layer contained 16 neurons. The network used for verification had used two neurons in the output layer that represented two classes of decision-acceptance or rejection of the signature. The training set contained 50 genuine examples per individual.

Different strategy was used in work in [50] where for each user a separate neural network was trained. When supplied with sample signature it decided whether the sample is genuine or forged.

In [51], fuzzy ARTMAP network was used for offline signature recognition. The authors' trained their system using only genuine signatures. This approach can be used when no examples of imitations are available in the learning set.

Examples of unsupervised learning approach can also be found in studies on handwritten signature recognition. In work [52] Abu-Rezq and Tobla used self-organizing maps to assess the inner and intra class variability of signature samples. Wessel and Omlin [53] investigated the application of Kohonen networks to cluster data in training set in order to construct signer models.

2.7.4.2.1.1 Hidden Markov Models   HMM proved to be a good classifier for handwritten text recognition. In particular, left-to-right HMM topology models the process of handwriting very well and is often applied to signature recognition. This topology was used by in [9] for offline signature verification. The last state of this model was connected to its entry to allow for verification of rotated signatures which were described by discrete radon transform projections. Another example using left-to-right topology is the work in [54]; for each user separate models with individual number of states were trained and used in multi expert setting where the outputs from differently parameterized models were combined to issue the final decision.

2.7.4.2.1.2 Support Vector Machines   SVMs classifiers were used for both [55] offline and online [56] signature classification. Typically, the polynomial and radial basis kernel functions methods are utilized. In [57] one can find a comparison of HMM and SVM classifiers applied to signature verification. Based on experiments, SVM is assessed to perform slightly better than the solution based on HMM.

2.7.4.2.1.3 Bayesian Networks   In work [58] Xiao and Leedham utilized Bayesian network for offline signature verification. The main advantage of Bayesian networks is the ability to model the variability among signatures given by the same person. The authors proposed experimental algorithm for construction of the network based on machine learning. The nodes of the network were divided into two classes—common hypotheses and alternative hypotheses to ensure that constructed network had a tree structure.

## 2.8 Assessment of Biometric Signature Systems

In order to evaluate signature biometric systems one can utilize standard measures that are used in other biometric traits, detailed description of these metrics can be found in [59].

In the identification task the success rate is used to assess system performance. It is computed as the ratio of properly identified signatures to total number of identification attempts.

For the verification task the most common measures are false acceptance rate (FAR), false rejection rate (FRR), and equal error rate (EER). FAR denotes percent of forged signatures that were accepted as genuine. It is computed as the ratio of number of falsely accepted forgeries to total number of verification attempts using signature imitations. FRR measures the percent of genuine signature that were incorrectly rejected as imitations.

As in other biometrics the decision to accept or reject given signature usually depends on the value of threshold that is imposed on the result of the system. FAR and FRR are functions of the threshold so their values are often presented as ROC (receiver operating characteristic) or DET (detection error trade-off) curves.

The selection of the threshold value influences the levels of both FAR and FRR errors. The change of the threshold that results in lowering one of these errors increases the other one. Due to this fact in many studies the systems are described in terms of EER which is obtained for the value of threshold when both FAR and FRR are equal.

## 2.9 Example Studies on Signature Recognition

### 2.9.1 Online System

In this section an online signature system is presented published by the authors in work [60]. To evaluate the system performance, a partial database is utilized from Signature Verification Competition (SVC2004), available for public use [10].

Signatures in this database were acquired by means tablet (WACOM Intuos). The functional features that were registered include pen coordinates $X$ and $Y$, pressure ($P$), elevation ($L$), and azimuth ($Z$). Online data collected with tablet device, in contrast to offline systems, do not require additional preprocessing such as localization and segmentation form background. However, some simple transformations such as resampling and normalization may improve the results.

Studies on dynamics of handwriting showed that the cut-off temporal frequency of the signing process is below 20 Hz [61]. This allows to reduce the input data by resampling (Nyquiest frequency requires only 40 samples per second) while retaining important components of writing parameters. In addition to this procedure another method—space-based sampling may be used. In this case, if the distance between consecutive samples is greater than heuristically selected threshold value, the second sample is omitted (Equation 2.17). With this method certain artifacts may be eliminated (e.g., when a writer holds a pen in one position at the end of writing).

$$d\big((x(t_i), y(t_i)),(x(t_{i+1}), y(t_{i+1}))\big) > T_{xy} \tag{2.17}$$

In this study the linear normalization was used to account for random translations and variations in signature scale. The results of these procedures are shown in Figure 2.25.

DTW is a method that models the time-axis fluctuation with nonlinear warping function [62]. It allows to compare two signals from the same source that have been affected by time-based distortions. The timing differences are eliminated by warping the characteristics of signals in such a way that the optimal alignment is achieved. In case of signatures the compared signals are sequences of samples ($X$, $Y$, $P$, $L$ and $Z$) after preprocessing. DTW algorithm defines a measure $D(A,B)$ between two sequences $A$ and $B$ (Equation 2.18):

$$A = < a_1, a_2, ..., a_M >$$
$$B = < b_1, b_2, ..., b_N > \tag{2.18}$$

| Before sampling | After sampling |
|---|---|

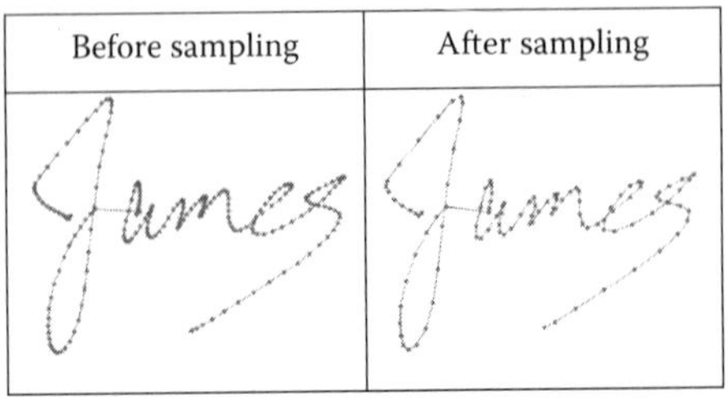

**Figure 2.25**   An example of online signature trajectory before and after sampling.

The distance $D$ is defined using the following Equations 2.19 and 2.20:

$$D(A,B) = D^R(N,M) \qquad (2.19)$$

$$D^R(i,j) = \min \left\{ \begin{array}{l} D^R(i,j-1) \\ D^R(i-1,j) \\ D^R(i-1,j-1) \end{array} \right\} + d(a_i,b_j) \qquad (2.20)$$

where $i = 1..M, j = 1..N$.

The distance $d(a_i,b_j)$ can be chosen in various ways depending on the application. In our case, the Manhattan distance is used. The calculations are carried out using dynamic programming with intermediate values of $D^R(i,j)$ stored in a cumulative matrix during computations (called cost matrix). Additional constraints such as window constraint [62] are added in order to prevent incorrect alignment.

In the basic version of DTW the partial distance $d$ do not incorporate any information on the context of values that are being matched. This may lead to incorrect results when features occurring in different contexts are treated as equivalent (e.g., local minima and maxima). In order to incorporate contextual information replace original distance measure $d(a_i,b_j)$ with $d_c(a_i,b_j)$ given by Equation 2.21:

$$d_c(a_i, b_j) = \frac{\sum_{k=-C}^{C} w_k d(a_{i+k}, b_{j+k})}{\sum_{k=-C}^{C} w_k}, \quad w_k = \frac{1}{1+\ln|k|} \qquad (2.21)$$

This measure uses weighted average over samples in the nearest neighborhood of points that are matched. The size of neighborhood is controlled by the value of $C$.

*2.9.1.1 Results*  Experiments were conducted using signatures from SVC2004 database [10] (genuine and skilled forgeries). Two modes of operation were investigated: writer identification and signature verification. In both scenarios one genuine signature per subject was used as a reference and the rest was utilized as a test set. The system was evaluated using signatures from 40 individuals (6 genuine and 20 skilled per subject). Experiments where repeated for every genuine sample used as a reference and the average values were computed for the percentage of properly identified writers (in identification mode) and EER (in verification mode).

*2.9.1.2 Identification*   In this scenario only genuine samples were used. For every subject one signature was selected as a reference and the rest was used for testing. Therefore, the reference set contained 40 signatures while the test set was composed of 200 samples. The evaluation was repeated for each genuine signature selected as a reference giving 48,000 comparisons (40 references samples × 200 test samples × 6). The average percent of properly identified signatures reached 99.2% when selected characteristics ($X, Y, L, Z$) were used.

*2.9.1.3 Verification*   In order to assess the presented method in verification mode to types of forgeries were used: random and skilled. Random forgeries are genuine signatures of individuals who try to use them to impersonate other signers. The reference set consisted of 40 samples (1 × 40), the test set for each subject was comprised of 5 genuine and 195 random forgeries (39 × 5). The lowest average EER = 0.83% was obtained when combined characteristics $X, Y$, and $P$ were used.

In the experiment with skilled forgeries the reference also contained 40 samples, the test set for each subject was formed from 5 genuine signatures and 20 skilled forgeries. The best result (EER = 5.99%) was achieved when all available features ($X, Y, P, L, Z$) where combined.

*2.9.1.4 Discussion*   The results in identification mode show that online registration of signing allows to create a system that has very good results, even in the case when only one reference sample is used. The level of error in the verification mode depends largely on the type of forgery and is significantly higher for skilled imitations. In this scenario the presented approach requires further investigation.

### 2.9.2 Offline System

This section presents an example of offline signature identification and verification system that was published by the authors in [5]. The experiments were carried out using signature images from GPDS database [11].

The signatures stored GPDS database are initially segmented and binarized so procedures such as signature localization and extraction are not required. The first stage of the proposed method reduced the amount of data to be processed. In order to achieve this goal the thinning procedure was applied and resulted in 1-pixel width signature lines that preserved complete topology of original shapes. There are many algorithms that may be used for this task, here we used our own method—KMM [63]. Figure 2.26 shows examples of original and thinned images by means of KMM method.

The signature features description and the distance measure used for comparison of signatures in this work was based on shape context method [38]. This method requires a signature representation as a set of points for which shape context descriptors can be calculated. The number of points that should be chosen depends on the complexity of object and in this work, based on previous experiments, the value of $N = 150$ points per signature were selected. One can choose various criteria for selection of points, however, in this investigation an approach suggested in work [38] was applied where selected points are equally distributed on the shape of graphical object. Examples of sampled signatures are showed in Figure 2.26.

As a result of this process each signature was initially described as a set of points (Equation 2.22):

$$A = \left\{ a_1, a_2, ..., a_i, ..., a_N \right\} \tag{2.22}$$

At the next step, the shape context descriptors were calculated. The shape context descriptors are log-polar space histograms computed

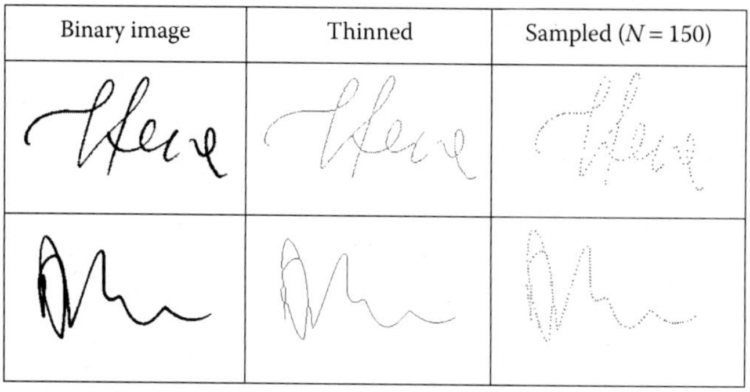

| Binary image | Thinned | Sampled ($N = 150$) |
|---|---|---|
| | | |
| | | |

**Figure 2.26** Examples of signatures images given by two individuals, their thinned and sampled versions.

separately for each point $a_i$ taken as the origin. The first step to this calculation is changing coordinate system from Cartesian to polar (Equations 2.23 and 2.24):

$$\theta_{ij} = \text{arctg}\left(\frac{y(a_j) - y(a_i)}{x(a_j) - x(a_i)}\right) \qquad (2.23)$$

$$r_{ij} = \log\left(\sqrt{\left[y(a_j) - y(a_i)\right]^2 + \left[x(a_j) - x(a_i)\right]^2}\right) \qquad (2.24)$$

where:

$x(a_i)$ is the X-coordinate of $a_i$

$y(a_i)$ is the Y-coordinate of $a_i$

Based on polar coordinates the bins of the histogram are determined (Equation 2.25):

$$h_k(a_i) = \#\left\{ \underset{j \neq i}{\forall} (r_{ij}, \theta_{ij}) : (r_{ij}, \theta_{ij}) \in \text{bin}(k)\right\} \qquad (2.25)$$

where bin($k$) is the $k$th bin of histogram describing the distribution of points in $A$ relative to point $a_i$, $K$ is the number of bins in the histogram (in our experiments we used $K = 60$). Finally, the signature image is represented as a set of histograms (shape context descriptors):

$$A_h = \{h(a_1), h(a_2), ..., h(a_i), ..., h(a_N)\} \qquad (2.26)$$

To compare two signatures $(A, B)$ represented using shape context descriptors $(A_h, B_h)$ a distance measure given in (Equation 2.27) was used:

$$D_h(A_h, B_h) = \frac{1}{N} \sum_{a_i \in A} \min_{b_j \in B} d_h(h(a_i), h(b_j)) \qquad (2.27)$$

where $d_h$ denotes cost of matching two histograms—one describing a point from object $A$ and the other from $B$, can be based on $\chi^2$ test statistics (Equation 2.28).

$$d(a_i, b_j) = \frac{1}{2} \sum_{k=1}^{N} \frac{\left[h_k(a_i) - h_k(b_j)\right]^2}{h_k(a_i) + h_k(b_j)} \qquad (2.28)$$

The total cost $D_h$ of matching two objects $A$ and $B$ that is used in this work is the sum of minimal distances between the points $a_i$ and $b_j$ given in (Equation 2.29).

$$D_h(A_h, B_h) = \frac{1}{N} \sum_{a_i \in A} \min_{b_j \in B} d_h(a_i, b_j)$$

(2.29)

The computation of histograms using basic approach is based only on points selected during sampling procedure and as a result describe their distributions. However, the selected points can be also utilized as the centers of histograms and the computation of the histogram can be based on original or thinned signature data (Figure 2.27). Such technique allows to incorporate more information about original object into descriptor. In this study we applied both techniques: the former was called "basic SC" and the latter named as "extended SC."

### 2.9.2.1 Results

The presented methods were assessed in two modes: identification and verification. In both scenarios the subset of GPDS database was used: eight examples of genuine signatures from 40 individuals (320 genuine samples in total). Half of this set (4 signatures per individual) was used as references while the other half was left for testing. In order to investigate the verification performance against skilled forgeries the subset of eight imitations per individual were selected at random from GPDS giving the total of 640 signatures in this experiment (320 genuine and 320 skilled forgeries).

### 2.9.2.2 Identification

In the identification the $k$-nearest neighbor classifier with $k = 4$ was applied with two distance measures based on "basic SC" and "extended SC." The experiments were repeated using

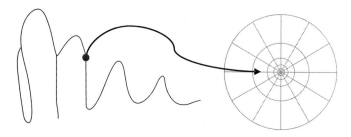

**Figure 2.27** Computation of histogram for selected point based on complete signature skeleton.

different reference and test sets and average values of percent of properly classified signatures were computed. The result obtained for the "basic SC" was 94%, the extended version reported 96% of properly classified signatures.

*2.9.2.3 Verification* During the verification task the questioned signature is compared with the references of a particular person to assess its authenticity. When the distance between the questioned signature and reference samples of a particular person exceeds the threshold value ($T$), the signature is rejected, otherwise it is recognized as a genuine example. In our experiment four reference samples were used which produced four distances when compared with the tested signature. The minimal distance value was used to make the final decision. For random forgeries the basic SC gave the average result of EER = 4.9%, while for the extended SC the EER was 4.4%. Figure 2.28 shows an example of the FAR and FRR curves obtained for extended SC.

In next series of experiments the system was tested against skilled forgeries, which are much more difficult to differentiate due their shape similarity to genuine signatures. In many cases such imitations can be almost impossible to detect in offline systems—the information contained in the static image may be not enough. In order to detect such forgery dynamic information on signing pressure and velocity is usually required that is available in online systems. The results obtained in this setting were EER = 22.6% ("basic SC") and EER = 20.4% ("extended SC").

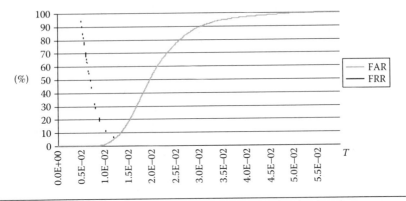

**Figure 2.28** FAR and FRR curves computed for extended SC and random forgeries.

*2.9.2.4 Discussion* The presented results show that the introduction of extended SC lowered the system error in all scenarios that were studied. Visual investigation of signature images from GPDS database and their thinned versions revealed that in several cases signature skeletons contained artifacts (e.g., tails and excessive loops). This could be the effect of binarization procedure that usually causes such problems. One of the solutions to address this issue could be to process the grayscale images directly, without thresholding the image to obtain binary map. This might improve the results even further.

It is also worth noticing that using extended context do not significantly increase the cost of computations (especially during identification mode). The computation time required by shape context method in order to compare two graphical objects increases with the number of points ($N$) used to describe the shape of those objects. The shape context for each signature is calculated only once (and stored for future usage), therefore do not introduce the need for additional computations during each comparison.

# References

1. D. Impedovo and G. Pirlo, Automatic signature verification: The state of the art. *IEEE Transactions on Systems, Man and Cybernetics, Part C: Applications and Reviews*, 38, 609–635, 2008.
2. A. J. Mauceri, *Feasibility Studies of Personal Identification by Signature Verification*. Space and Information Systems Division, North American Aviation Co., Anaheim, CA, 1965.
3. R. Plamondon and G. Lorette, Automatic signature verification and writer identification. *Pattern Recognition*, 22, 107–131, 1989.
4. A. K. Jain, A. Ross, and S. Prabhakar, An introduction to biometric recognition. *IEEE Transactions on Circuits and Systems for Video Technology*, 14, 4–20, 2004.
5. M. Adamski, K. Saeed, M. Tabedzki, and M. Rybnik, Signature system based on extended shape context descriptors. In *International Conference on Biometrics and Kansei Engineering*, IEEE, Tokyo, Japan, pp. 267–272, 2013.
6. L. S. Olivera, E. J. R. Justino, C. Freitas, and R. Sabourin, The graphology applied to signature verification. In *12th Conference of the International Graphonomics Society*, Salerno, Italy, pp. 286–290, 2005.
7. J. K. Guo, D. Doermann, and A. Rosenfeld, Off-line skilled forgery detection using stroke and sub-stroke properties. In *15th International Conference on Pattern Recognition*, IEEE, Barcelona, Spain, pp. 355–358, 2000.

8. F. Zoebisch and C. Vielhauer, A test tool to support brute-force online and offline signature forgery tests on mobile devices. In *Proceedings of the International Conference on Multimedia and Expo*, IEEE, Baltimore, MD, pp. 225–228, 2003.

9. J. Coetzer, B. M. Herbst, and J. A. du Preez, Offline signature verification using the discrete radon transform and a hidden Markov model. *EURASIP Journal on Applied Signal Processing*, 2004, 559–571, 2004.

10. D.-Y. Yeung, H. Chang, X. Yimin, S. George, R. Kashi, T. Matsumoto, G. Rigoll, SVC2004: First international signature verification competition. In *International Conference on Biometric Authentication*, Springer, Hong Kong, pp. 16–22, 2004.

11. J. F. Vargas, M. A. Ferrer, C. M. Travieso, and J. B. Alonso, Offline handwritten signature GPDS-960 corpus. In *9th International Conference on Document Analysis and Recognition*, IEEE, Paraná, Brazil, pp. 764–768, 2007.

12. J. Ortega-Garcia, J. Fierrez-Aguilar, D. Simon, J. Gonzalez, M. Faundez-Zanuy, V. Espinosa, A. Satue et al., MCYT baseline corpus: A bimodal biometric database. *IEEE Proceedings of Vision, Image and Signal Processing*, 150, 395–401, 2003.

13. S. Garcia-Salicetti, C. Beumier, G. Chollet, B. Dorizzi, J. Leroux les Jardins, J. Lunter, Y. Ni et al., Biomet: A multimodal person authentication database including face, voice, fingerprint, hand and signature modalities. *Audio- and Video-Based Biometric Person Authentication*, Lecture Notes in Computer Science, 2688, 845–853, 2003.

14. H. D. Crane and J. S. Ostrem, Automatic signature verification using a three-axis force-sensitive pen. *IEEE Transactions on Systems, Man and Cybernetics*, SMC-13, 329–337, 1983.

15. M. E. Munich and P. Perona, Visual identification by signature tracking. *IEEE Transactions on Pattern Analysis and Machine Intelligence*, 25, 200–217, 2003.

16. I. Yoshimura and M. Yoshimura, Off-line verification of Japanese signatures after elimination of background patterns. *Progress in Automatic Signature Verification*, 13, 693–708, 1994.

17. S. Djeziri, F. Nouboud, and R. Plamondon, Extraction of signatures from check background based on a filiformity criterion. *IEEE Transactions on Image Processing*, 7, 1425–1438, 1998.

18. A. Chalechale, G. Naghdy, P. Premaratne, and A. Mertins, Document image analysis and verification using cursive signature. In *IEEE International Conference on Multimedia and Expo*, IEEE, Taipei, China, pp. 887–890, 2004.

19. H. Srinivasan and S. Srihari, Signature-based retrieval of scanned documents using conditional random fields. In *Computational Methods for Counterterrorism*, S. Argamon and N. Howard (eds.), Springer, Heidelberg, Germany, pp. 17–32, 2009.

20. M. Martinez-Diaz, J. Fierrez, M. R. Freire, and J. Ortega-Garcia, On the effects of sampling rate and interpolation in HMM-based dynamic signature verification. In *9th International Conference on Document Analysis and Recognition*, IEEE, Paraná, Brazil, pp. 1113–1117, 2007.

21. M. Adamski and K. Saeed, Online signature classification and its verification system. In *7th Computer Information Systems and Industrial Management Applications*, IEEE, Ostrava, Czech Republic, pp. 189–194, 2008.

22. R. Doroz, P. Porwik, T. Para, and K. Wróbel, Dynamic signature recognition based on velocity changes of some features. *International Journal of Biometrics*, 1, 47–62, 2008.

23. E. M. Nel, J. A. Preez, and B. M. Herbst, Estimating the pen trajectories of static signatures using hidden Markov models. *IEEE Transactions on Pattern Analysis and Machine Intelligence*, 27, 1733–1746, 2005.

24. S. Lee and J. C. Pan, Offline tracing and representation of signatures. *IEEE Transactions on Systems, Man and Cybernetics*, 22, 755–771, 1992.

25. K. K. Lau, P. C. Yuen, and Y. Y. Tang, Stroke extraction and stroke sequence estimation on signatures. In *16th International Conference on Pattern Recognition*, IEEE, Quebec, Canada, pp. 119–122, 2002.

26. Q. Yu, L. Jianzhuang, and T. Xiaoou, Offline signature verification using online handwriting registration. In *IEEE Conference on Computer Vision and Pattern Recognition*, IEEE, Minneapolis, MN, pp. 1–8, 2007.

27. G. Dimauro, S. Impedovo, G. Pirlo, and A. Salzo, A multi-expert signature verification system for bankcheck processing. *International Journal of Pattern Recognition and Artificial Intelligence*, 11, 827–844, 1997.

28. J. J. Brault and R. Plamondon, Segmenting handwritten signatures at their perceptually important points. *IEEE Transactions on Pattern Analysis and Machine Intelligence*, 15, 953–957, 1993.

29. Y. Rekik, N. Houmani, M. A. El Yacoubi, S. Garcia-Salicetti, and B. Dorizzi, A comparison of feature extraction approaches for offline signature verification. In *International Conference on Multimedia Computing and Systems*, IEEE, Ouarzazate, Morocco, pp. 1–6, 2011.

30. C. Santos, E. J. R. Justino, F. Bortolozzi, and R. Sabourin, An offline signature verification method based on the questioned document expert's approach and a neural network classifier. In *Proceedings of 9th International Workshop on Frontiers in Handwriting Recognition*, IEEE, Tokyo, Japan, pp. 498–502, 2004.

31. L. Schomaker, From handwriting analysis to pen-computer applications. *Electronics & Communication Engineering Journal*, 10, 93–102, 1998.

32. T. A. Osman, M. Krishnan, and M. J. Paulik, A minimum-velocity-based segmentation scheme for improved performance of an online signature verification system. In *IEEE International Conference on Electro/ Information Technology*, IEEE, Ames, IA, pp. 78–83, 2008.

33. W.-S. Lee, N. Mohankrishnan, and M. J. Paulik, Improved segmentation through dynamic time warping for signature verification using a

neural network classifier. In *International Conference on Image Processing*, IEEE, Chicago, IL, pp. 929–933, 1998.

34. T. H. Rhee, S. J. Cho, and J. H. Kim, On-line signature verification using model-guided segmentation and discriminative feature selection for skilled forgeries. In *Proceedings of Sixth International Conference on Document Analysis and Recognition*, IEEE, Seattle, WA, pp. 645–649, 2001.

35. L. L. Lee, T. Berger, and E. Aviczer, Reliable online human signature verification systems. *IEEE Transactions on Pattern Analysis and Machine Intelligence*, 18, 643–647, 1996.

36. J. Galbally, J. Fierrez, M. R. Freire, and J. Ortega-Garcia, Feature selection based on genetic algorithms for on-line signature verification. In *IEEE Workshop on Automatic Identification Advanced Technologies*, IEEE, Alghero, Italy, pp. 198–203, 2007.

37. M. Adamski and K. Saeed, Offline signature verification based on shape contexts using shared and user-specific thresholds. *Journal of Medical Informatics and Technologies*, 22, 195–201, 2013.

38. G. Mori, S. Belongie, and J. Malik, Efficient shape matching using shape contexts. *IEEE Transactions on Pattern Analysis and Machine Intelligence*, 27, 1832–1837, 2005.

39. K. Saeed and M. Adamski, Experimental algorithm for characteristic points evaluation in static images of signatures. In *Biometrics, Computer Security Systems and Artificial Intelligence Applications*, K. Saeed, J. Pejas, and R. Mosdorf (eds.), Springer, New York, pp. 89–98, 2006.

40. B. Fang, C. H. Leung, Y. Y. Tang, K. W. Tse, P. C. K. Kwok, and Y. K. Wong, Off-line signature verification by the tracking of feature and stroke positions. *Pattern Recognition*, 36, 91–101, 2003.

41. T. Kaewkongka, K. Chamnongthai, and B. Thipakorn, Off-line signature recognition using parameterized Hough transform. In *Proceedings of the 5th International Symposium on Signal Processing and Its Applications*, IEEE, Brisbane, Australia, pp. 451–454, 1999.

42. M. A. Ferrer, F. Vargas, C. M. Travieso, and J. B. Alonso, Signature verification using local directional pattern (LDP). In *IEEE International Carnahan Conference on Security Technology*, IEEE, San Jose, CA, pp. 336–340, 2010.

43. Y. Qi and B. R. Hunt, Verification of handwritten signature images by multiresolution wavelet analysis. In *Signals, Systems and Computers*, IEEE, Pacific Grove, CA, pp. 6–10, 1993.

44. Z. Ma, X. Zeng, L. Zhang, M. Li, and C. Zhou, A novel off-line signature verification based on adaptive multi-resolution wavelet zero-crossing and one-class-one-network. In *Advances in Neural Networks – ISNN 2007*, Lecture Notes in Computer Science, D. Liu, S. Fei, Z.-G. Hou, H. Zhang, C. Sun (eds.), vol. 4493, Springer, Heidelberg, Germany, pp. 1077–1086, 2007.

45. H. Lei, S. Palla, and V. Govindaraju, ER2: An intuitive similarity measure for on-line signature verification. In *9th International Workshop on Frontiers in Handwriting Recognition*, IEEE, Tokyo, Japan, pp. 191–195, 2004.

46. Q.-Z. Wu, S.-Y. Lee, and I.-C. Jou, On-line signature verification based on logarithmic spectrum. *Pattern Recognition*, 31, 1865–1871, 1988.

47. Y. Mizukami, K. Tadamura, M. Yoshimura, and I. Yoshimura, Statistical displacement analysis for handwriting verification. In *Image Analysis and Processing—ICIAP 2005*, Lecture Notes in Computer Science, F. Roli and S. Vitulano (eds.), vol. 3617, Springer, Heidelberg, Germany, pp. 1174–1181, 2005.

48. K. Huang and H. Yan, Off-line signature verification using structural feature correspondence. *Pattern Recognition*, 35, 2467–2477, 2002.

49. N. Mohankrishnan, W.-S. Lee, and M. J. Paulik, Multi-layer neural network classification of on-line signatures. In *IEEE 39th Midwest Symposium on Circuits and Systems*, IEEE, Ames, IA, pp. 831–834, 1996.

50. D. Z. Lejtman and S. E. George, On-line handwritten signature verification using wavelets and back-propagation neural networks. In *6th International Conference on Document Analysis and Recognition*, IEEE, Seattle, WA, pp. 992–996, 2001.

51. N. A. Murshed, F. Bortolozzi, and R. Sabourin, Off-line signature verification using fuzzy ARTMAP neural network. *IEEE International Conference on Neural Networks*, vol. 4, IEEE, Perth, WA, pp. 2179–2184, 1995.

52. A. N. Abu-Rezq and A. S. Tolba, Cooperative self-organizing maps for consistency checking and signature verification. *Digital Signal Processing*, 9, 107–119, 1999.

53. T. Wessels and C. W. Omlin, A hybrid system for signature verification. In *International Joint Conference on Neural Networks*, IEEE, Como, Italy, pp. 509–514, 2000.

54. J. R. Riba, A. Carnicer, S. Vallmitjana, and I. Juvells, Methods for invariant signature classification. In *15th International Conference on Pattern Recognition*, IEEE, Barcelona, Spain, 953–956, 2000.

55. M. A. Ferrer, J. B. Alonso, and C. M. Travieso, Offline geometric parameters for automatic signature verification using fixed-point arithmetic. *IEEE Transactions on Pattern Analysis and Machine Intelligence*, 27, 993–997, 2005.

56. A. Kholmatov and B. Yanikoglu, Identity authentication using improved on-line signature verification method. *Pattern Recognition Letters*, 26, 2400–2408, 2005.

57. E. J. R. Justino, F. Bortolozzi, and R. Sabourin, A comparison of SVM and HMM classifiers in the off-line signature verification. *Pattern Recognition Letters*, 26, 1377–1385, 2005.

58. X. Xiao and G. Leedham, Signature verification using a modified Bayesian network. *Pattern Recognition*, 35, 983–995, 2002.

59. R. M. Bolle, J. Connell, S. Pankanti, N. K. Ratha, and A. W. Senior, *Guide to Biometrics*, Springer, New York, 2004.

60. M. Adamski and S. Khalid, Signature verification by only single genuine sample in offline and online systems. In *International Conference of Numerical Analysis and Applied Mathematics*, Rhodes, Greece, 2015 (in press).

61. G. Lorette and R. Plamondon, Dynamic approaches to handwritten signature verification. In *Computer Processing of Handwriting*, R. Plamondon and C. G. Leedham (eds.), World Scientific, pp. 21–47, 1990.

62. H. Sakoe and S. Chiba, Dynamic programming algorithm optimization for spoken word recognition. *IEEE Transactions on Acoustics, Speech and Signal Processing*, 26, 43–49, 1978.

63. K. Saeed, M. Tabędzki, M. Rybnik, and M. Adamski, K3M—A universal algorithm for image skeletonization and a review of thinning techniques. *International Journal of Applied Mathematics and Computer Science*, 20, 317–335, 2010.

# 3

# KEYSTROKE DYNAMICS

Keystroke dynamics (also keystroke biometrics, typing dynamics, keyboard dynamics, keystroke analysis, typing biometrics, and typing rhythms) is the detailed information that describes the keystrokes as a person is typing at a keyboard or similar hardware. The definition above is an extension of the definition given in [1]. A keystroke is the event of pressing and releasing a single key. Keystroke dynamics is a measurement and a description of multiple keystrokes. Keystroke dynamics is based on the assumption that different people type in uniquely characteristic manners. Keystroke dynamics belongs to biometrics (in particular, behavioral biometrics), that is widely used in computer security [2]. Behavioral biometrics (how a human behaves) is generally in opposition to physiological biometrics (how a human looks like); however, hand signature [3] is considered as belonging to both. Behavioral biometrics patterns are statistical in nature. Examples of other behavioral features are voice and gait [4]. There are many other physiological biometrics that are currently researched. Biometric methods are used mostly for authentication and identification. In the past few decades, there was a noticeable increase in biometrics popularity, especially in the domain of data security. It is believed, however, that behavioral biometrics is not as reliable as physiological biometrics (with many very reliable physiological biometrics as fingerprints or DNA).

## 3.1 History of Keystroke Dynamics

The history of keystroke dynamics is dated as far as the invention of the telegraph and its popularization in the 1860s. A telegraph message was sent using Morse code by pressing a key rhythmically.

The telegraph operators got so experienced in sending "dot" and "dash" signals that they were recognizable by their coworkers based on their way of sending messages. This method was called "the Fist of the Sender" and it turned out to be very helpful during the World War II where it allowed allied forces to verify the authenticity of the messages. The next era began when personal computers were popularized. With the ease of data gathering and processing complemented with a large amount of different keys (originated from typewriters), the full potential of keystroke dynamics was quickly revealed in password hardening and increasing security of computer systems. This allowed for signal analysis and user authentication using keystroke dynamics.

There have been several scientific studies on using keystroke dynamics for user verification [4–9]. Most studies have used durations between keystrokes as features for user verification, but some have also used keystroke durations (the time a key is held down).

## 3.2 Keystroke Analysis

The basic information acquired by keystroke dynamics is

- When a key was pressed
- When the key was released
- Which key was pressed

The devices used to capture keystroke dynamics may be, for example,

- Standard physical keyboards
- Phone keyboards or keypads (ATM)
- Specialized physical keyboards (with better time-resolution [10] or force-sensitive [11])
- Touchscreen keyboards (where keyboard is displayed on touchscreen, in lieu of being physical device [12,13])
- Touchscreen keyboards with "swype" (an input manner for touchscreen that allows for drawing a curve through the keys corresponding to a word, rather than tapping the individual keys)
- TV remote keyboards

The next section details the data acquisition procedures.

### 3.2.1 Data Acquisition

Traditionally, a keystroke dynamics system consists of two phases: (1) enrollment, when users keystroke dynamics is registered and memorized (user profile is created) and (2) comparing new data to the profiles collected. Phase 1 consists of writing some text, it may be the same text for all users (easing profile comparisons), individual text for each user (e.g., secret phrase—password), or unspecified text. The data registered in profile may vary depending on the classification method used.

The keystroke rhythm of a user is measured and encoded in a way to develop the unique template for future authentication [14]. Raw measurements available from standard keyboard (usually also processed with operating system [OS] message queue) can determine two basic types of events (MS Windows standard is given as example, being the most popular general purpose OS):

WM_KEYDOWN (keycode $k$)—when the key $k$ is pressed down

WM_KEYUP (keycode $k$)—when the key $k$ is released

Exemplary keystrokes for the typed text "litwo! ojcz" are presented in Figure 3.1 as a set of binary functions; one binary function for

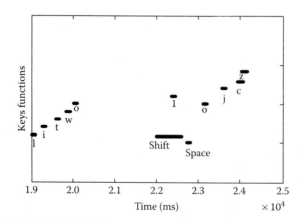

**Figure 3.1** Keystrokes versus time (please notice the use of "shift" to produce "!" character and overlapping of the last two keystrokes "c" and "z"). (Data from M. Rybnik et al., A keystroke dynamics based system for user identification, *IEEE-CISIM08 Computer Information Systems and Industrial Management Applications*, Ostrava, Czech Republic, 2008. With permission.)

each possible key (values {0;1}, 0 for nonpressed state, 1 for pressed state) versus time. Note that such data representation allows to register overlapping, that is usually supported by hardware, up to several simultaneous keystrokes. Computer keyboards typically allows for overlapping (called also anti-ghosting due to missing keystrokes effect) of 4–10 keys, depending on the purpose and affordability.

The timing data is then processed to determine a pattern for the future comparison [16]. The methods for creating the user pattern differ significantly, ranging from the simplest statistics, like the number of keystrokes per minute, to storing each individual keystroke as well as keystroke successions in pairs, triplets, and so on.

The most popular derivative features obtainable from "timing data" are as follows:

- *Dwell time*: The time a key is in the pressed state.
- *Flight time*: The time between a key is released and the next key is pressed.
- *Typing speed*: Average number of keystrokes per time interval.
- *Overlapping of specific keys combinations*: Especially "shift" or "Caps lock" for writing capital/small letters, but also overlapping of letters predicated by fast typing.
- *Key preferences*: Selecting text before or deleting letters one by one, as well as a manner of typing used for corrections that may be very distinct as in most cases only one key will be pressed, however some users may place the cursor in approximate middle of text to be deleted and use both "delete" or "backspace" keys simultaneously, then delete the remaining letters using only one key.
- *Usage of specific keys*: like arrows, page up, page down, home, end, shift, control, keypad, and so on.

More advanced domain- and hardware-dependent information may be acquired, such as

- What was the force during the pressure (related to the time between start and end of a keystroke requires a special keyboard [17]).
- Amount of errors (practically: how often a user uses "delete" or "backspace" keys).

- Pressing finger's size and relative position on key, as well as movement (obtainable for touchscreen visual keyboard [12,13]).
- Video feed of hands during typing.

Flight and dwell are popular features because they can be easily measured with standard PC hardware or even using webpage with JavaScript application embedded [18]. Key events ("press down" and "release") events generate hardware interrupts. Gathering keystroke dynamics data is not straightforward, as two or even several keys can be pressed at the same time—when a person presses next key(s) before releasing the previous one(s)—it happens often when typing fast, or using special keys (Shift, Alt, Ctrl). Flight times might then be negative, what increases processing complexity [19]. A typical scenario of user enrollment, with a fixed text for all users is presented in Figure 3.2.

The event timing is affected by internal clock. In [20], keyboard functioning using 15 MHz function and arbitrary waveform generator was examined. It was noticed that 18.7% of keystrokes were

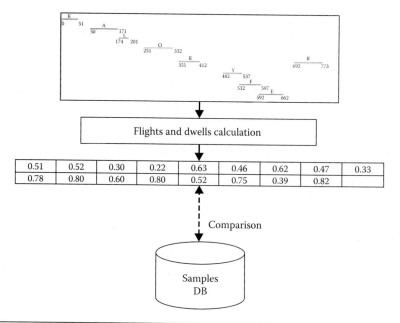

**Figure 3.2** Typical scenario for user enrollment.

acquired with a 200 µs error. Therefore, the keyboard was calibrated and the database was collected using higher precision. Data has been gathered with the accuracy of 100 µs. This experiment has shown that databases gathered using different machines may not be comparable because of "the lack of the main bus clock calibration." What is more, when the CPU(s) is under load, the delays in keystroke acquisition occur as they are usually handled by the message queue (in Windows these are the messages *WM_KEYDOWN* and *WM_KEYUP*). There is also a difference between the lengths of keyboard clock frames of popular operating systems: Linux/Unix (10 ms frame) and Windows (15 ms frame, 64 ticks per second). The influence of time resolution on the algorithm classification accuracy should be taken into account.

In most applications (disregarding keystroke dynamics) the information when a key was pressed or released is not stored, neither when typing in a local (desktop computer) nor remote environment (typing a text on a webpage, using remote console). The only stored/sent data is a text (in details: a sequence of characters from a specific subset, like ASCII, Unicode, etc., without any timestamp). The process of acquiring keystrokes is called keystroke logging. Keystroke logging usually captures much more information than a sequence of characters (text), a character may need one, two, or even more keystrokes to be inserted into text. Capital letters need shift and a letter key; diacritic-modified letters may need alt and a letter key, special characters (e.g., "!," "@," "#", "$," and "%"), and so on may need shift and a number key, tilde "~" may need three keystrokes: simultaneous shift and tilde key with successive space key, and so on. Each keystroke may be described with many features, including primary: key, time of pressing down, time of releasing, force of pressing, force of aftertouch; and derivative: dwell. The delay between two keystrokes and the time between releasing the previous one and pressing the next one (flight) is another derivative information. Additionally keystrokes may overlap, this is required for the above-mentioned multikeys characters, but also occurs during fast typing. In Table 3.1 there is an example of typical characters and keystrokes needed to type them.

Statistics (and models) can be calculated either without regard to the key, that is, combined for all keys; or with regard to the key (gathered for every keystroke separately—allowing establishing models for each key or chording separately). Even statistics for approaches

**Table 3.1**  An Example of Characters and Keystrokes Needed

| CHARACTER | NUMBER OF KEYSTROKES | KEYSTROKES |
|---|---|---|
| "c" | 1 | <"c", 232, 286> |
| "C" | 2 | <"shift", 358, 789><"c", 402, 590> |
| "!" | 2 | <"shift", 892, 1056><"1", 941, 1079> |
| "ą" | 2 | <"left alt", 1243, 1677><"a", 1389, 1505> ("Polish programmers" keyboard layout) |
| "~" | 3 | <"shift", 1697, 1886><"~", 1776, 1912> <"space", 2234, 2469> |

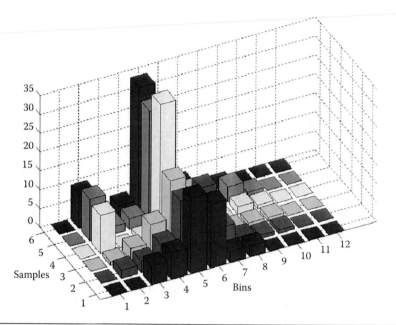

**Figure 3.3**  Six statistics for "dwell" feature of two different individuals (samples 1–3 and 4–6) without regard to the key. One can observe similar functions distribution for the two users. (Data from M. Rybnik et al., A keystroke dynamics based system for user identification, *IEEE-CISIM08 Computer Information Systems and Industrial Management Applications*, Ostrava, Czech Republic, 2008. With permission.)

without regard to the key may be significant [15], as could be seen in Figure 3.3.

In [21], there is an example of statistics (average and standard deviations) for word "password" (approach "fixed text for all users") typed several times by three different persons. There are clear differences

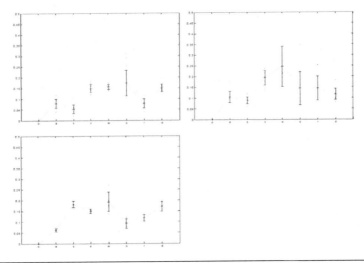

**Figure 3.4** The statistics for word "password" (average and standard deviations) typed several times by three different persons. (Data from J. Ilonen, Keystroke dynamics, *Advanced Topics in Information Processing–Lecture 03–04*, Lappeenranta University of Technology, Lappeenranta, Finland, 2002. With permission.)

that may be used for establishing model and verifying the user identity. It is important to note that such model will be viable for typing "fixed text," and not just any text (Figure 3.4).

### 3.3 Variability of Users, User Behavior, and Hardware

As with the other biometrics—especially behavioral ones—there are several factors increasing the variability of keystroke dynamics data. One problem is that there is a wide variety of typing skills. The speed of typing alone can be very different between different typists. An experienced typist writes much faster than a beginner. The predictability of a fast writer is much greater—as he is more proficient (and uses mostly "muscle memory" to type well-known words)—while the keystroke dynamics for a beginner may be rather random (and easy to imitate).

The behavioral biometrics in general may be heavily affected by user attitude, state, and circumstances. The typing is affected by user apparatus condition (contusions), alertness (sleepy, hyperactive, and well-rested), and background activity like talking, thinking what to type next, drinking coffee, and so on.

A change in typing hardware (like using different keyboard than usual or using a laptop computer instead of a desktop PC) can be seen as a noise, which heavily affects keystroke dynamics. Another source of problems is the use of software that imitates typing: macros generator, a virtual keyboard, pasting from clipboard, and a voice-to-text converter.

There are successful commercial products that are able to handle these issues and have proven effective in large-scale use (thousands of users) in real-world settings and applications; however, their range of applications is still limited.

### 3.4 Authentication and Identification

One of the most common and important tasks in data and systems security is authentication (identity verification)—determining if the specific persons (or the remote computer in the network) identity conforms to its own claim. Regarding a person, the process can be seen as the confirmation of the given identity. After successful authentication the user is allowed to access the data or services of the system.

*Slightly more difficult task is identification*: Determination of the user's identity, without any claim. In this case the whole database has to be searched and the matching user may be authenticated and given access.

Keystroke dynamics may be used in various scenarios:

1. *Support for user authorization* [22,23]: When a user accesses the system, he/she is required to type a text, further analyzed by keystroke dynamics in order to confirm his/her claimed identity. In this case keystroke dynamics measurements are compared with only one previously stored profile. The disadvantage of such a procedure is rather small amount of keystrokes that are collected (for user convenience) and possible changes in keystroke dynamics as the user continues to type (and therefore practices) the same text for a long time. Generally, keystroke dynamics is not considered reliable enough to be used for user authorization alone at a large scale or forensics (like fingerprints or signature, that have huge tradition and reputation). It may be used however to rule out some people from being the analyzed typist.

2. *Password hardening method*: An additional authentication mechanism, used in parallel to login/password authentication.

Exemplary scenario of password hardening method is security algorithms verify at first if the typed password is correct, and at the next stage, the keystrokes sample is being analyzed by keystroke dynamics algorithm. Therefore, even if the password is correct but it is not typed in the right way, the user may be denied access to the system. Keystroke dynamics blends very well with such login/password authentication, however one should regard legal issues and OS mechanisms that may serve to bypass keystroke dynamics authentication (such as telnet access, pasting passwords from clipboard, etc.). Another disadvantage is password security precautions that may require periodical changes of passwords (and therefore require another lengthy keystroke dynamics data enrollment). An example of such system is AdmitOneSecurity [24] (formerly BioPassword).

3. *On-the-fly user authorization*: During the work of the user already logged to the system the keystroke dynamics data are collected, analyzed in comparison to user keystrokes profile, and preventive activities may be taken in case of doubts on user identity. The procedure is more viable, as it allows gathering large keystroke data (assuming the user is actively typing of course) and it does not prevent the user from work unless significant dispersions are found between his current behavior and stored profile. Suspicious activity (considering keystroke dynamics) may be a reason for additional security measures or checks. This could prevent "password sharing" (a user that reveals his personal password to another), leaving the computer system logged-in that enables another person to use it or authorization under a direct threat from an attacker.

4. *Remote access for all above*: Keystroke dynamics has significant advantages in "remote access" (especially with mobile use—access from different places). Most of the authentication methods based on tokens (as keycards) are not convenient to use, as there is no possibility to check if the authorized person uses the token; and tokens frequently require specific hardware. The authentication is in most cases done using passwords or masked passwords, where the user inputs only some random part of the password (thus in case of security breach protecting the remaining part of the password and

thus preventing the fraudulent use), that suffer from *password sharing* threat. This scenario calls for biometric solution, but a significant disadvantage of most biometric authorization or verification techniques for distributed systems is the need for specialized input hardware, like cameras, microphones, finger scanners, hand (fingers and palm) scanners, drawing tablets, signature tablets, and others. Such hardware may be hardly available at a remote workstation assuming the access to a secure system may be granted from many places. Keystroke dynamics however is obtainable using a standard computer keyboard, which is normally accessible at remote station and a web-based application (e.g., website with JavaScript technology [25]). Keystroke dynamics may be used without the need for a specialized hardware, so assuming that communication protocol timestamps (with high accuracy up to tenths of milliseconds) each keystroke may be used for access verification to remote systems. Reproducing biometric input may potentially be detected, as biometric data (from the same individual) are never exactly the same. Thus, using keystroke dynamics authorization techniques is interesting because of the lack of hardware requirements (contrary to the most biometric techniques) and keeping the main advantage of biometric techniques: uniqueness (at least up to some point).

5. *Identification* [16]: With enough keystroke data collected, keystroke dynamics analysis may be performed in order to link the data with data previously collected for a specific group of system users. Based on the analysis the user's identity is deducted. It requires a large amount of keystroke dynamics data collected and therefore is not a very practical application.

*3.4.1 On Biometrics Context of Keystroke Dynamics*

The most commonly used classification of human authentication methods was introduced by Wood in [26]. With addition of biometrics one can distinguish three groups of user authentication methods:

- Proof by knowledge (something that the user remembers), such as passwords, PIN numbers, lock combinations, secret questions

- Proof by possession (a kind of token that the user possesses), for example, keys, chip cards, magnetic cards, hardware, or software tokens
- Biometrics (physiological body characteristics or the behavior of the user), such as fingerprints, signature, eye retina pattern, hand shape, keystroke dynamics, and so on

Proof by knowledge is the most popular method of securing digital data, commonly referred to as password. For high security, it is important to follow three rules listed in [27]: complexity, uniqueness, and secrecy. Unfortunately, in reality most people ignore at least one of the rules, for example:

1. The password is unique and kept secret, but simple to guess.
2. The password is unique and complex, but written on a piece of paper.
3. The password is complex and secret but identical for every service.

As reported by Schneier [28], about 25% of the passwords can be guessed using a 1000-word dictionary with 100 common suffixes. Combining this with biographical data and using a bigger dictionary brings success rate up to 55%–65%. This certainly points out the main weakness of proof by knowledge techniques: human convenience.

Techniques that use proof by possession do not guarantee high security nor availability. As the tokens are physical objects, they can be possibly handed over, stolen, misplaced, or broken. They need (and usually they are) to be secured by an additional password or a PIN code (proof by possession and proof by knowledge). Unfortunately, such additional code is frequently used in public and therefore its secrecy is seriously threatened.

The closest correspondence to keystroke dynamics among biometrics is online signature recognition. Both online signature recognition and keystroke dynamics try to identify people by their writing (typing) dynamics, which are assumed to be unique (up to some point) among different people. A much more popular biometrics: offline signature recognition is however more similar to image recognition and is quite popular application among biometrics. In online signature recognition however not only the resulting signature is examined but the whole process of signing: how the person signs, including the speed, pen pressure, and changing angulate of pen in two axes.

Reliability of a biometric system is measured by rates of type I error (false rejection rate [FRR]; also false negative) and type II error (false acceptance rate [FAR]; also false positive). FRR means rejection of a valid user and FAR means the acceptance of an invalid user. In the first case, a valid user gets angry because he could not use the system or is required to repeat the authentication procedure. In the second case, a user without proper authorization (possibly an attacker) is allowed to access the system. Both error rates should be optimally 0%, what is rather improbable. Usually there is a natural balance between the two error types: when one kind of error is made lower by tuning the system, the other usually increases. As an example, if all users are accepted—the biometric authorization is basically turned off—FRR will be 0% but FAR may be very high as all potential attackers are allowed into the system. From a security point of view FAR should be minimized—as it means no chance for an illegal user to access the system. However, FRR should be also low because valid users get annoyed if the system makes their lives harder. Practical applications are tuned regarding these two error rates, depending if they are more focused on security or the users' comfort. Behavioral biometrics uses a confidence measurement instead of the traditional pass/fail measurements. The traditional benchmarks of FAR and FRR no longer have linear relationships. Keystroke dynamics (as well as the other behavioral biometrics) can adjust FRR/FAR at the individual level. This allows for individual risk stating—a thing physiological biometrics do not achieve so easily.

## 3.5 Characteristics of Keystroke Dynamics

In his work, Jain [29] presents an extensive comparison of various biometric techniques. He proposes to evaluate the efficiency of biometrics purposes in terms of the following parameters:

- *Universality*: Describes how commonly the biometric is found individually. A case when the particular biometric feature cannot be obtained from the user, as it may sometimes happen (a simple example is finger blessing altering the fingerprint);

thus, systems based on more than one feature (so-called multimodal) are desirable; as described by [24] where voice, hand
geometry, and face image are used together.

- *Uniqueness*: How well the biometric separates individually
  from another.
- *Permanence*: Measures how well the biometric resists aging.
- *Collectability*: Defines ease of measurements acquisition.
- *Performance accuracy, speed, and robustness*: Practical aspects of
  using the specific biometric technology.
- *Acceptability*: Degree of approval for the technology.
- *Circumvention*: Defines the ease of producing a substitute; in
  other words: how easy it is to falsify the specific biometric.

Furthermore, Jain [29] proposes a table that presents evaluation of these
parameters for biometric techniques in the following scale: H = High,
M = Medium, and L = Low. The evaluation is presented in Table 3.2,
which was given in Chapter 2 and is presented here for convenience.

### 3.5.1 Universality

Regarding the modern Western countries, keyboard typing is common, but in broad context many people either do not use (older gen-
erations) or cannot even have access to computers (poor countries with
little access to electronics or even electricity). Therefore, the universality of keystroke dynamics as compared to other biometrics is really
low, as most physiological biometrics, gait, and even signature is more
common than proficiency in typing. Therefore, the universality of keystroke dynamics is rather limited, as it is not uncommon to encounter
people, whose keystroke dynamics profile cannot be registered.

### 3.5.2 Uniqueness

Keystrokes are rated "low" by Jain [29] in the matter of uniqueness, what probably is caused by both (1) the total inability of some
people to type and (2) low typing proficiency among most people,
that do not use computer and typing machines frequently enough.
Low typing proficiency influences both uniqueness and classification
accuracy.

**Table 3.2** Comparison of Signature with Other Biometric Traits

| BIOMETRIC IDENTIFIER | UNIVERSALITY | DISTINCTIVENESS | PERMANENCE | COLLECTABILITY | PERFORMANCE | ACCEPTABILITY | CIRCUMVENTION |
|---|---|---|---|---|---|---|---|
| *Signature* | L | L | L | H | L | H | H |
| Face | H | L | M | H | L | H | H |
| Fingerprint | M | H | H | M | H | M | M |
| Hand geometry | M | M | M | H | M | M | M |
| Keystrokes | L | L | L | M | L | M | M |
| Hand veins | M | M | M | M | M | M | L |
| Iris | H | H | H | L | H | L | L |
| Retinal scan | H | H | M | M | H | L | L |
| Voice | M | L | L | M | L | H | H |
| Facial thermogram | H | H | L | H | M | H | L |
| DNA | H | H | H | L | H | L | L |
| Gait | M | L | L | H | L | H | M |
| Ear | M | M | H | M | M | H | M |

*Source:* A.K. Jain et al., *IEEE Trans. Circuit Syst. Video Technol.,* 14, 4–20, 2004. With permission.

### 3.5.3 Permanence

The permanence of keystroke dynamics is rated low as typing may be seen as a skill, that changes gradually (sometimes even rapidly) in time. It means that the individual typing profile is changing in time, following the changes in proficiency with keyboard, health state, and other issues. Particularly, striking changes occur when measuring keystroke dynamics with the use of a fixed text, as typing proficiency of well-known phrases raises rapidly. One can however prevent the loss of precision by continuous updating of the stored profile.

### 3.5.4 Collectability

One can see that keystrokes are rated middle in the matter of collectability; this is caused by the fact that in order to collect keystrokes one needs specialized software installed in individual machine. However, this is compensated by the fact that the hardware required to gather the data (keyboard) is (in simple case) standard computer system equipment, contrary to most other biometric techniques that require additional, often expensive, equipment.

### 3.5.5 Performance

Keystroke dynamics performance depends highly on the used processing algorithms, however generally it is regarded as medium, probably due to slight complication of the raw data, which is similar to multivariate time series. Time dependency makes the processing intrinsically complicated; usually some simplifications and statistics are used instead of analyzing raw data.

### 3.5.6 Acceptability

The work of Li et al. [25] points out that the researchers often overlook an important disadvantage of many biometric methods: acceptability. Obtaining fingerprints or an iris scan may be considered insulting or scary by some people. The acceptability of the technique is ranked by A. K. Jain as middle, suggesting that people are not easily offended by that form of biometric data gathering. Keystroke dynamics is a non-invasive method, and this is a major advantage. The system analyzing the keystroke dynamics can work without notifying the user and it can be supported by other biometric methods in multibiometric systems. It is important however to be aware of legal dangers of using keystroke monitoring software or hardware, that may be seen as keylogging and therefore limited or forbidden by local law.

### 3.5.7 Circumvention

Circumvention of keystroke dynamics data is ranked medium, as in most cases it is possible to falsify biometric data, possessing the biometric profile used for comparison or original raw data sequence. In the first case, an attacker in possession of biometric profile may generate keystrokes data matching the features examined. In the second case, original raw data sequence may be falsified by slightly changing the data, in order to avoid the unlikely perfect match of biometric (assuming the data is stored somewhere in raw format). With such knowledge an attacker may take possession of false (but matching!) keystrokes and depending on the accessibility to communication channel reproduce the keystroke sequence. This is especially dangerous when acquisition is performed over the web, what is more difficult to control physically.

### 3.5.8 Summary

Regarding these rankings, it is important to properly define the potential uses of keystroke dynamics. With a close examination of the above-mentioned properties, keystroke dynamics is not reliable enough to be a stand-alone method, where strict security is expected. Currently it is mostly used along a traditional password as an additional authentication mechanism, for example, as password hardening method.

Keystroke dynamics, like signature and gait analysis, has significant advantages over other biometrics. It is noninvasive, highly acceptable, and in its basic form it does not need specialized hardware. Tokens or the biometric systems can be used to strengthen security, but they require expensive additional hardware and/or user time. Contrastingly, keystroke dynamics acquisition and analysis may be transparent to the user.

There are also some disadvantages of keystroke biometrics:

- Efficient interpretation of features can be problematic.
- Limitations of operating systems can affect the data quality.
- Database creation and quality determination for keystroke dynamics is hardly defined. These subjects are detailed in [12].

Another use of keystroke dynamics is related to work time monitoring. An employer can verify if the employee is working as he is supposed to, or even if someone else is working on his computer. However, this application is controversial since employees may feel offended and not being trusted. The use of such application in most cases requires consent of the employee. Yet another problem is the storage and transmission of acquired keystrokes data and profiles, as it may be used for attack on secrecy of data processed by employee (e.g., business or industrial secrets).

There is some disagreement about the future of the authentication methods. In 2011 IBM published a report entitled: "The Next 5 in 5 – Innovations that will change out lives in the next 5 years" [30]. As the authors predict, we will no longer use traditional passwords, which will be replaced by the biometric methods in the next 5 years. On the other hand, research division of Microsoft published a paper [31] in which they argue that the passwords are here to stay for the foreseeable future and they also list numerous problems with them. In these circumstances, it is worth considering if using keystroke dynamics does not present a reasonable alternative solution on this matter.

## 3.6 Approaches to Keystroke Dynamics

This section is aimed at categorization of approaches and presenting examples of the most popular configurations and tasks related to keystroke dynamics.

### 3.6.1 Taxonomies of Approaches

Keystroke dynamics may be used for various tasks, most important are specified in Section 3.4. It is viable also to properly define the input and processing of keystrokes, in order to make the obtained results comparable.

### 3.6.2 Input Text Approach Taxonomy

Regarding the "input text approach" keystroke dynamics can be divided into four variants:

1. *Fixed text for all users*: A text is strictly specified and every user has to type it during the enrollment. Usually this involves also correcting of eventual errors (disregarding of the whole sample), when the user makes a mistake in the specified text. Profile matching and examining is simplified, as every user performs the same exercise, therefore the results are directly comparable.

2. *Fixed text for each user*: A text is specified for every user individually and the user has to type it during the enrollment. Usually this involves also correcting of eventual errors (disregarding of the whole sample), when the user makes a mistake in the specified text. This variant is used in "password hardening" application (see Section 3.4). Profile matching and examining is simplified, as usually every user performs the same exercise, and therefore the results are directly comparable [32].

3. *Non-fixed text with regard to key* (differentiation between keys): The typed text is not specified, but there is a regard to which key is pressed/released, sometimes sequences of keys are also analyzed (sequences—even the shortest possible: pairs— require longer samples, as at least part of them should appear both during enrolment and identity verification).

4. *Non-fixed text and no regard to key* (no differentiation between keys): The typed text is not specified, and there is a regard to which key is pressed/released. Potentially it is not as informative variant; however such globally stated statistics are also proven to provide some interesting results [15].

There are many approaches with the technology used ranging from simple statistical measures to artificial intelligence techniques.

### 3.6.3 Simple Typing Features

*Keystrokes per minute*: May serve to eliminate a part of keystroke profiles, being unlikely that someone may suddenly type twice as fast than previously. The opposite case may be however possible, therefore the use for this simple statistics is limited. It may however serve as an additional sample feature.

*Characteristic typing manner*: Every user has some customs while typing, some of them may be very characteristic and serve to differentiate between keystroke profiles. For example, not using keypad for entering numbers, using left/right shift with some particular letters [10], pressing both delete and backspace simultaneously while deleting larger words, not using jumping to next/previous words, not using page up/page down, and so on. While such manners may not always be present, detecting such characteristic may be used as additional information.

*Right- and left-handiness*: It is estimated that handiness influences also typing, what may serve for a hint to differentiate profiles. Right-handed people may be faster in getting to keys they press with right-hand fingers than with left-hand fingers, simply because their particular hand is more dexterous.

*Native/fluent language influence*: In English, the words "the," "be," "to," "of," and "and" are the most common [32], and those words for English-proficient typists are well-known sequences that are typed much faster than other keys in similar order and position on keyboard. Similarly common prefixes (such as "dis," "ir," "re," and "un") and suffixes (such as "-ing," "-ed," "-ly," "-es") [33] may be entered faster than similar sequences (regarding key locations on keyboard) of no particular meaning.

*Key overlapping*: When typing really fast or well-known words, it is characteristic that slight overlapping of specific keystrokes might occur, as the keys are pressed in very quick succession, like "arpeggio" [34] in music. It is obviously related to native/fluent language influence, as the typed phrases should be well-known to the typist.

*Frequent errors*: Common "typos" may be characteristic for a person, and there are many types of errors, such as substitutions, adjacent letter mishits, drop-outs, reversals, double-strikes, quasi-homonyms, and extra-hold errors (i.e., shift key held down too long or too short). These errors may be detected with dictionary check, but also by detecting the corrections (that may be not easy for a keystroke-monitoring application, as corrections may be made also with mouse). The errors profile might be used as feature in keystroke sample classification.

*Individual statistics for specific keys*: The time to get to and depress a particular key (flight), and the time the key is held-down (dwell) may be very characteristic for a person, regardless of the overall typing speed. This is directly related to the location of keys in the keyboard, and fingering of the keys. Another important factor is the succession of keys in typed text that affects the fingering. Please note that the keyboard standard "qwerty" was made with a thought to facilitate typing for English language. However, it wasn't supposed to make the work easier. In fact the goal was exactly the opposite. Keys in this layout are placed in such locations that the statistically most common sequences of letters in English language are being separated. This was meant to reduce the frequency of moving typebars internal clashing.

*Fingers dexterity*: Fingers have different characteristics, and their use with particular keys varies by person. Generally, some fingers are faster and more dexterous than other; resulting in greater typing speed and probably also preferred. Particularly striking example is a beginner two index fingers typing technique, which is slow and characteristic.

*Individual statistics for specific successions of keys*: Highly influenced by the knowledge of a specific keys succession (and more generally words, that is correlated to the knowledge of specific language). In addition, some sequences of letters may occur frequently for given language/domain and therefore be characteristic. It is important to mention the huge influence of the user proficiency with the particular keyboard layout, as well as the intrinsic typing speed and properties of keyboard layout itself.

## 3.7 Advanced Approaches

A number of approaches have been researched for authentication based on keystroke dynamics data [35–37]. The methods used to process keystroke dynamics data may be, for example,

- *Statistical techniques*: The most simple approach is to measure latencies between consecutive key presses (flights) or key pressing latencies (dwells) and calculating their statistical properties (e.g., average and standard deviation). The statistics are stored in the user profile. Depending on the input text variant for fixed text repetitions might be required; for non-fixed text a large sample may be needed. An example of a profile for fixed text (including average and standard deviation for each keystroke) was presented in Figure 3.1. Such profile could be used for direct comparison with a sample taken. Additionally after positive authentication, the profile could be updated with the sample just taken, what could help with profile permanence.
- *Minimum distance*: The most simple features comparison method, with the correct feature selection and weighting often gives good preliminary indications, such as
  - The Bayes rule [38]
  - ARTMAP [39]
- Decision trees [40].
- *Pattern matching (DTW)*: Advanced pattern comparison method, advantageous because compared patterns may be of different length. For keystroke dynamics it may mean that the keystroke sequences with errors may be also compared, with no need for intervention.
- Hidden Markov models [41].
- *Artificial neural networks* (ANNs; e.g., RBF neural networks, LVQ neural networks [42]): They are the common tools for classification methods nowadays. Frequently they are following of modifying statistical methods (e.g., clustering and Bayes classifier). The main disadvantage of ANN is the high dependence on the training database and high cost of retraining. After training however ANNs are frequently quick to use, and also, usually they are black-box models, so no information about the inside mechanics is available.

A comprehensive exploration of different classification methods used for keystroke dynamics is given in [6].

## 3.8 Fixed Text for All Users

Fixed text for all users variant is probably the most efficient group of approaches as it allows for direct keystroke-to-keystroke comparison. The example application below was published by the authors in [43]. The authors have tested the approach on a large group of individuals (about 250), with data gathered using browser-based WWW JavaScript application. The samples analyzed were of different length, ranging from 9 to 54. Samples were preprocessed to eliminate errors and unnecessary special keystrokes.

### 3.8.1 Dataset

The keystrokes database has been gathered with WWW application located at site http://home.agh.edu.pl/~panasiuk/kds/. Collected raw data consists of scan code of every keystroke, information whether key was pressed or released, time of event in milliseconds measured from the first keystroke. Please note that web browser (as well as user's OS and hardware/software configuration) can add noise to raw data. Website users had to register themselves once and then they can log in and type samples. We have gathered over 1100 samples in the database, left by over 250 registered individuals. A sample consists of five different phrases from the same user. The phrases are selected in specific way

- A native Polish tongue twister: *"Stół z powyłamywanymi nogami"*
- A short Polish word *"kaloryfer"*
- A common sentence in Polish
- A sentence in English—foreign language for majority of users, *"After some consideration, I think the right answer is."*
- An user-chosen phrase (usable in "fixed text for each user variant")

Some declarative information was obtained for each individual: declared proficiency with typing (beginner, average, advanced), gender and whether he/she is left/right-handed. Further in this section authentication experiments on the second fixed text input: one-word phrase "kaloryfer" are presented.

### 3.8.2 Proposed Algorithm

The keystroke features extracted were dwell and flight, defined respectively as time when the specific key is in pressed state, and time between releasing one key and pressing the next. The flight may be also negative if overlapping occurs. Every sample was preprocessed by cleaning from special keys like "backspace," "shift," "alt," or cursor keys, in order to obtain a homogenous sample (what is however a significant data modification, as obtaining flight after deleting keystrokes is usually impossible). As the aim of the study is to directly compare fixed texts, typing errors were also disregarded, along with corresponding flight time between removed letters.

We have implemented a simple algorithm, examining corresponding dwell and flight times when comparing the matching keystrokes for the same text. The symbol represented by keystroke (letter, number, etc.) and position in phrase is taken into account, what makes the data much more "informative," however simultaneously limiting practical applications. Classification of samples is based on $k$-nearest neighbor algorithm.

"A sample" is defined here as a vector containing correct dwells and flights for given fixed text (as some could be incorrect and removed). For the analyzed word "kaloryfer" it contains nine dwells and eight flights.

Data normalization has been performed by counting maximum and minimum "dwell" and "flight" values for specified phrase in whole database (all users and samples). Normalized flight $f_{norm}$ have been counted as follows:

$$f_{norm} = \frac{f - f_{min}}{f_{max} - f_{min}}$$

and normalized dwell $d_{norm}$ in analogical way:

$$d_{norm} = \frac{d - d_{min}}{d_{max} - d_{min}}$$

The "distance" $d$ between two samples was calculated as a weighted sum of *dwells-distance* $d_d$ and *flights-distance* $d_f$:

$$d = p \cdot d_f + (1 - p) \cdot d_d$$

where the parameter $p$ changes in range $[0; 1]$ and signifies the importance of "flights-distance" in relation to "dwells-distance."

*Dwells-distance $d_d$* was measured in the following way:

$$d_d = \sum_{n}^{i=1} |d_{1i} - d_{2i}|$$

where $d_{1i}$ and $d_{2i}$ are $i$th valid corresponding normalized "dwells" for samples 1 and 2.

*Flights-distance $d_f$* were measured in the following way:

$$d_f = \sum_{m}^{i=1} |f_{1i} - f_{2i}|$$

where $f_{1i}$ and $f_{2i}$ are normalized flight times of $m$ valid corresponding flights in phrase for samples 1 and 2.

We have randomly selected a "Reference Database" (known samples to be compared with unseen samples) and "Generalization Database" (to be classified). In first experiments the Reference Database consisted of $k$ (ranging from 1 to 9) samples from each user, while the rest of samples were put into Generalization Database. We have also performed experiments with constant number of samples in the Reference Database.

The classification strategy contains a voting mechanism. $k$ closest distances are calculated and sorted from the best vote (1st) to the worst one ($k$th). Then we assign *token count $t_i$* for each vote as follows:

$$t_i = k - i + 1$$

where $i$ is the rank on closest distance list.

In this way the shortest distance gets token count equal to $k$, the second one gets $k - 1$, and so on, finally the $k$th closest distance gets token count equal to 1. If two or more votes were cast for the same user then their "token counts" are counted and the user with the highest token count "wins"—his ID is being returned as a result.

Below we can see the experimental results for one-word phrase "kaloryfer," with a variable number of samples in the Reference Database (increasing with $k$).

In the Figure 3.2 we can see correct classification rate (percentage) for different values of parameter $p$ (from 0 to 1 in 0.1 steps) and various $k$ nearest neighbors taken into account (from 1 to 9) (Figure 3.5).

As one can see parameter $p$ (flight time importance) has significant influence on the results. The best results are obtained for the parameter in the range of [0.4; 0.6] what means almost equal importance of both features. Extreme values of parameter (0 and 1) where either only flights or only dwells are taken into account are clearly worse than the medium range.

One can also note significant increase of classification accuracies with increase of number of neighbors $k$ taken into account. It is related to the fact that more samples are included in the reference database for higher values of $k$ and therefore a study with a constant number of samples in the reference database seems to be necessary.

Figure 3.6 presents the results for constant number of users and samples in the Reference Database.

This study shows that the voting strategy significantly increases the classification accuracies, from 54% for winner-takes-all ($k = 1$) to 70% ($k = 9$). That justifies the use of the voting mechanism. Constant number of samples in the Reference Database has, however, definitively increased the competing chances for lower number of neighbors, as their results are much better than before. Therefore it is clear that more samples in the Reference Database results in better classification.

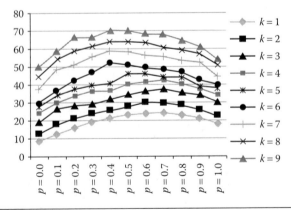

**Figure 3.5** Classification accuracies (% of correct user matches) for phrase "kaloryfer," depending on dwell/flight importance parameter $p$ and number of nearest neighbors $k$.

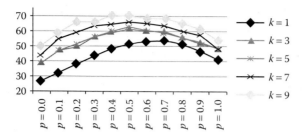

**Figure 3.6**    Classification accuracies (% of correct user matches) for phrase "kaloryfer," for 19 users and with the constant number of samples in the "Reference Database."

**Table 3.3**    Best Classification Accuracies for Phrases

| PHRASE | NUMBER OF KEYSTROKES IN PHRASE | BEST CLASSIFICATION ACCURACY (%) |
|---|---|---|
| "kaloryfer" | 9 | 70.37 |
| "After some consideration, I think the right answer is:" | 54 | 88.55 |
| "Stół z powyłamywanymi nogi" | 28 | 92.5 |

We have performed similar experiments for two other phrases from gathered database: <foreign language (English) phrase> "After some consideration, I think the right answer is:" and a native language (Polish) tongue twister "*Stół z powyłamywanymi nogami.*" The overall tendencies in classification accuracies were very similar, with the best results obtained for "flight/dwell weight" close to 0.5 area and for nine samples in the Reference Database (maximal number researched). One can deduct that researching and regarding more neighbors results in better classification accuracy. Table 3.3 shows the best results obtained for the three phrases researched.

Examining the table one can notice that longer phrase allows to obtain significantly better classification accuracy. One can note also that <foreign language phrase> however longer proved to be slightly more difficult to classify than <tongue twister>. We suspect that it comes from the fact that typing in foreign language is much more difficult for most users, therefore they type slower so the resulting samples fail to carry as much of the individual typing habits as for the native language. It is however a hypothesis that require further study.

It is possible to achieve very promising classification rates over a considerably large group of individuals with the use of quite low number of keystrokes, when comparing the matching keystrokes for the same text. Used procedure of comparing only matching keystrokes somehow limits the usefulness of the variant (e.g., impossible to use in password hardening) and also requires pre-processing to eliminate typing errors.

The fixed text for each user variant was tested on a group of people with regard of the possible use of this authentication technique as a support for user authentication over the Internet. With analysis of only two keystrokes features and with the use of relatively simple classification techniques the keystroke dynamics proved to be promising and effective biometrics for identification/authentication of individuals. It is necessary to stress that with use of constant text it is possible to effectively distinguish a vast majority of users with a relatively short keystrokes sequence (beginning from nine keystrokes). It is necessary to stress the advantages: possibility to use a booming Internet as a natural transmission medium for the biometrics and the lack of need for dedicated acquisition devices. That clearly shows the potential of keystroke biometrics for the authentication of individuals over the Internet.

Possible improvements:

- Use of more advanced classification algorithms
- Integration of more keystroke metrics

### 3.9 Fixed Text for Each User (BioPassword/AdmitOneSecurity)

#### 3.9.1 Computer-Access Security Systems Using Keystroke Dynamics

One of the most known studies in the area is "Computer-access security systems using keystroke dynamics" by Bleha et al. [6]. In this study keystrokes were extracted only from users' usernames. The identity of the user was determined based on the analysis of the manner they wrote their usernames using latencies between keystrokes. The database contained 30 latest valid username entries. That allowed the system to successfully accommodate to gradual changes in user typing manner. Two classification methods were used: minimum distance and Bayesian classifier.

#### 3.9.2 AdmitOneSecurity

AdmitOneSecurity [24] (formerly BioPassword) offers a patented commercial system that uses keystroke dynamics for password strengthening what according to the proposed taxonomy is equal to fixed text for each user variant. Keystroke dynamics authentication serves here to confirm the user's identity and detect unauthorized access for example resulting from password sharing.

The proposed solution replaces the default log-in system used in Windows operating systems, especially working in a local network. In networked environment accounts are stored centrally on a server. To implement additional accounts authorization with keystroke dynamics password strengthening. Dedicated software has to be installed on the central server, as well as on all client workstations. During user enrollment, he is required to type his login and password many times (by default 15). Biometric templates are determined and stored centrally. During log-in the system checks not only correctness of the login attempt against the stored template. Only a user whose typing pattern matches the stored template is allowed to log in.

## 3.10 Non-Fixed Text with Regard to Key

Non-fixed text with regard to key variant is less efficient approach than fixed text variants, but it is very flexible as it allows for text-independent comparison. Potentially this also allows for continuous authentication, as described in Section 3.12. The example application described below was published by the authors in Ref. [44] and is aimed at efficient user authentication with keystroke dynamics using non-fixed text of various lengths.

Samples were not error-eliminated as it is virtually impossible to detect errors (except for "backspace" and "delete" keystrokes, that may be misleading). We have tested the approach on a small group of individuals, with data gathered in various ways: over Internet using browser-based WWW application and on local machines using dedicated application. The keystrokes are aggregated depending on the key pressed, and the statistics of such distribution are analyzed.

### 3.10.1 Proposed Algorithm

Keystrokes samples were collected from nine individuals (which corresponds to the conditions of small office or department). Each user typed a long text (of more than 250 characters) twice in the five sessions, to provide 10 samples for each person. Experiments were carried out as follows for given person keystroke sample was chosen randomly (in some studies, it was the whole sample, but most used the shorter fragment to simulate the usage for shorter texts) and an attempt was made to identify or verify that individual identity. Therefore, the text might be seen as non-fixed one, as keystroke-to-keystroke comparison is not performed, even though it is sometimes possible. The keystrokes were aggregated instead, depending on the key pressed. The remaining data is treated as the Knowledge Database (i.e., training set). The test was repeated thousand times and obtained results were averaged. The rates of correct identification and verification (percentage of samples correctly identified or verified) are shown in the graphs and tables below.

On the basis of the training set the following model of each of the subjects were created. For each key separately average and variance of dwell and flight times were calculated. The variance was used for normalization in part of experiments. In order to perform identification, it was only necessary to compare it with all registered user models (for authentication with only one user), and not all samples collected in the training set. This allowed speeding up the process without compromising the results. Dwell distance $d_d$ without normalization was obtained in the following way:

$$d_d = \sum_{c='a'}^{'z'} \left| d_c^{\text{test}} - d_c^{\text{train}} \right|$$

where $d_c^{\text{train}}$ and $d_c^{\text{test}}$ are the dwell times of $c$ letter, for "train" and "test" samples respectively.

Flight distance $d_f$ without normalization was obtained in the following way:

$$d_f = \sum_{c='a'}^{'z'} \left| f_c^{\text{test}} - f_c^{\text{train}} \right|$$

where $f_c^{\text{train}}$ and $f_c^{\text{test}}$ are the flight times of $c$ letter, for train and test samples, respectively.

A sort of data normalization has been performed in some experiments using dividing by standard deviation for dwell and flight values. The standard deviation was calculated for all samples corresponding to a specific letter.

Therefore *normalized dwell distance* $d_{dn}$ was obtained in the following way:

$$d_{dn} = \sum_{c='a'}^{'z'} \frac{\left| d_c^{\ test} - d_c^{\ train} \right|}{\sigma_{dc}}$$

where $d_c^{train}$ and $d_c^{test}$ are the dwell times of $c$ letter, for train and test samples respectively.

*Normalized flight distance* $d_{fn}$ was obtained in the following way:

$$d_{fn} = \sum_{c='a'}^{'z'} \frac{\left| f_c^{test} - f_c^{train} \right|}{\sigma_{fc}}$$

where $f_c^{train}$ and $f_c^{test}$ are the flight times of $c$ letter, for train and test samples respectively.

In the same manner the test sample was created. The average dwell time and flight time for each typed character were calculated, and these values formed a feature vector. The feature vector was compared to the model of a person, by measuring distance between them, taking into account only the characters present in the test sample (some characters might not be present at all particularly for short text). The distance was calculated according to the taxicab metric (Manhattan distance). Finally the distances were classified using variants of nearest neighbor classifiers. It is worth noting that the two compared text parts were selected randomly, and most likely were different.

### 3.10.2 Experimental Results and Discussion

The first set of experiments was designed to examine the impact of the entered text length on the recognition efficiency. The test was repeated for different number of characters in entered sample from 8 (which is equal to the common length of a password) to 250 (corresponding to longer statements of about 50 words in English). The obtained results are shown in Table 3.4.

As one can see, texts a mere 100-character long already managed to provide decent identification accuracy of more than 90%. Use of longer texts leads to further improvement. This indicates that testing the text entered during normal user operation, can allow the identification of the writer with a high efficiency. The system may be

**Table 3.4** Identification Results for Various Text Lengths

| SAMPLE LENGTH (STROKES) | CLASSIFICATION ACCURACY (%) |
| --- | --- |
| 8 | 50.5 |
| 15 | 62.1 |
| 30 | 72.4 |
| 50 | 82.4 |
| 100 | 91.6 |
| 150 | 92.8 |
| 200 | 93.2 |
| 250 | 97.2 |

therefore considered as a promising tool to ensure the safety of both: home computer and remote access terminal. In subsequent studies, the authors tested various conditions.

The first change concerned the treatment of the test sample instead of calculation of statistics for the entered characters; one can treat the given string as a feature vector, and compare it with the model generated from the training set.

In addition, it was examined what impact on the classification outcome will have exclusion of flight time from tests (this value is subject to significant fluctuations during writing longer texts, such as when the user hesitates what to write next, therefore its impact can be confusing and deteriorate the results).

The authors also examined how the results change when the sample is not normalized with nearest neighbor classifier. One would expect that this will result in a deterioration of the results; however, in the case of the existence of outliers, the calculated variance may be too high and cause incorrect scaling. This study shows the effect of different lengths of the sample as well as the other parameters.

Another experiment shows the results of the user authentication problem the problem here was not to identify a person, but to confirm (or reject) its identity. Any user requesting access to the system declares his presumed identity and the system compares the features collected from the entered text with a corresponding model of a person. The study includes both the correct input (when the user enters his authentic data), as well as cases of fraud (when the user is impersonating someone else). This allows the calculation of two error rates: FAR and FRR. Usually the most interesting value is EER (equal error rate, when both errors are equal).

The study was conducted on 200-characters long texts (from earlier experiments it was concluded that this value is sufficient for the proper recognition of a person), counting statistical properties and using the variance, but without the flight time.

The EER value obtained was 3.1%, which is a very promising result. The relationship of the values of FRR and FAR is worth mentioning. In a system with a high level of security it is possible to select such threshold, that FAR value is less than 0.1% (which results in a lower probability of fraud than a guess at four-digit PIN) and the FRR will be around 50%, what means that every other time the system will require a double-entry of data (or the introduction of an additional 200 characters).

A non-fixed text with regard to the key variant was tested for the possible use of identification and a support for user authentication. With analysis of only two keystroke features and with the use of simple classifier the keystroke dynamics proved to be a promising and effective biometrics feature for individual authentication. It is necessary to stress that with the use of non-fixed text it is possible to effectively distinguish a vast majority of users with a relatively short keystrokes sequence. Even more interesting is the use of on-the-fly user authentication while the user is performing normal interaction with the system—involving longer texts, alike what the experiments showed—a minimum of 200 keystrokes. Short user authentication consisting of several keystrokes seems to be too short for the practical use.

Possible improvements:

- More advanced classifiers as well as integration (eventually development) of more keystroke metrics. The features could include among others: specific character overlapping and linking keystrokes features to keystroke sequences in time. This possibly requires longer reference samples, but would result probably in an increase of the classification accuracy.
- Verification of possibility and efficiency of integrating the keystroke dynamics information into secure communication protocols. This could be achieved by keystrokes timestamp attachment to fragile data, when someone would like to confirm the identity of the author. It might be considered as a biometric variation of a signature.

## 3.11 Non-Fixed Text with No Regard to Key

Non-fixed text with no regard to key variant is probably the least accurate, compared the other variants. However one can imagine situations, where information about what key was pressed is inaccessible or insecure to provide. One possible scenario is obtaining keystroke-like information from audio recording of a typing person. Depending on the keyboard, there may be characteristic sounds of pressing and releasing keys, what may be also individual for the key (what slightly exceeds keystroke dynamics). This setup however would require above-mentioned variant to be applied. The example application below was published by the authors in [15] and is aimed at efficient user authentication with keystroke dynamics using non-fixed text with no regard to key.

### 3.11.1 Dataset

The keystroke input was taken from a group of 37 individuals. Learning and validation was performed on completely different texts of short size (110 keystrokes [first and second samples] used as reference and about 55 [third sample] keystrokes used for validation). The texts were independent of each other as they did not contain any repetitive words. For the sake of algorithm we have tested only the "letter" keystrokes and "space," ignoring "shift," cursor keys, "delete," "backspace," and other nonalphanumerical keys, thus obtaining rather similar keystrokes, without any keys that might be of prolonged use (like keeping "backspace" pressed in order to delete a mistakenly typed word).

The keystroke features extracted were dwell and flight, defined respectively as time when the specific key is in pressed state, and time between releasing one key and pressing the next. The flight may be also negative if overlapping occurs. By eliminating "shift" and "alt" we have eliminated common overlapping resulting from writing capital letters using "shift" and using "alt" for diacritical marks. As such we have obtained only overlapping occurring from fast typing, what we have qualified as distinctive feature—helpful in classification.

According to the operating system documentation the time of keystroke message may vary with an error of 10–15 ms. However, it is not sure whether the error occurs or not, so it is still interesting to measure the keystrokes within more detailed time intervals.

### 3.11.2 Proposed Algorithm

In order to quantify the data representing time we have decided to split it into multiple "bins" for easier perception. In Section 3.2.1, one can see an example of "dwell" feature, put into 12 bins (Figure 3.3). This example presents 6 samples, where samples 1, 2, 3 and 4, 5, 6 are from two different users.

The number and boundaries of bins were stated globally for the whole database and determined in empirical way. The values of "dwell" were distributed from 0 to 140 ms uniformly within 2 ms interval. The values of "flight" were distributed uniformly from –20 to 475 ms within 2 ms interval. All values outside the boundaries were put into the closest extreme bin, thus eliminating the need for elimination of outlying observations.

We have performed two independent experiments, first taking into account only "dwell," then using only data with "flight," and finally using the two features in the specific way. In all experiments we have used a very simple classification measure—closest Manhattan distance between referring keystroke feature vector and the feature vector to be classified. We have classified the verification vector to have the same class as the closest of referring vectors.

First approach based only on dwell provided a mere 64.86% correct classification rate. The second approach based on flight provided 53.76% correct classification rate. For the third approach we have combined the two features. We have calculated the distances matrices separately and added them applying weight equal to the fraction of arithmetic means for the distance (calculated over the whole matrix). In this way, we have obtained 72.97% of correct classification rate.

As one can see with this simple variant and no regard to the key it is possible to achieve near 73% of correct classification over a group of 37 individuals with homogenous data and low number of keystrokes. That shows the great potential of keystroke biometrics even with such obscure data available.

Next, we have modified the classification strategy by introducing a voting mechanism (of three closest distances). In the situation where second and third distance pointed for the same candidate, a measurement was taken of the degree of importance of the votes in order to verify if they were close or distant comparing with the first distance.

It was realized by using weights signifying tolerance: 0.9 and 0.8 for the second and third closest distance, respectively. If these distances modified by weights are on average better than the "winner," then the decision is based on their conclusion, which has to be the same for both. That allowed us to take profit from such situations resulting in a significant increase of the overall classification accuracy, which finally was equal to 72.68% for both dwell and flight data taken simultaneously into account.

### 3.12 Continuous Authentication

Continuous authentication is a widely used scenario that has been implemented for numerous commercial applications. Shepherd in [45] presents a study of keystroke dynamics used for continuous authentication scheme. The main problem with such application are anomalies, that need to be accounted for (e.g., typing with only one hand, sleepiness or other psychic states, keyboard change, hand wounds, etc.). On the other hand, in such application usually huge amount of data is available, that makes processing easier. What is more: after positive authentication, further enrollment may be conducted, which allows for compensation of gradual changes (especially increase of typing proficiency). As confirmed by this study forcibly logging out the user was too disruptive and the system rather generates warning for human administrator, than takes excessive actions on its own. The authors had promising results using only flights and dwells and calculating the averages and variances globally. These very simple features were good enough for authentication of rather small number of four users.

### 3.13 Perspectives

Latest research focuses in general on the user authentication in order to secure personal computers. There are only a few works on the topic of user identification. Scientists working on keystroke dynamics could pay more attention to recognizing the most important features, correlations, and factors affecting users' characteristics.

With many of the ideas on algorithms tested, researches started looking for new features that would improve the accuracy. The work regarding errors made by users and their correction methods (backspace or delete key) is [10]. There was no error analysis, but the users preference of employing the left or the right shift key was tested leading to a slight improvement in the accuracy. Hence, adding new characteristic features can improve the accuracy of the keystroke dynamics algorithms. Naturally, using specialized hardware results in additional features. One of the ideas is to use pressure-sensitive keyboards. Microsoft is working on the hardware [46] and a student team contest was organized using the prototypes, searching for new ideas [47]. It is shown that pressure is even a more important characteristic than the dynamics itself [48]. In [39] the authors constructed their own keyboard and used pressure as an additional feature, which turned out to be very helpful for the user authentication. These results suggest that the use of pressure feature would greatly help in keystroke dynamics analysis. The main problem is however the lack of pressure-sensitive keyboards.

Mobile phone keyboards are interesting due to the rising popularity of devices and specific requirements regarding their use (quick access and high comfort) [12]. Keystroke dynamics may be used to authenticate users in-the-flight. Their predecessor—standard phone keyboard—was also researched and the error rates are reported to be about 12.8% for 11-digit phone number [49], 13% using fixed 10-character long alphabetical input [50].

The application for ATM hardware rather than keystroke dynamics used keystroke motion and hand shape at different time points [52]. Achieved error rates were as low as 1.1%–2.7% depending on the PIN and exact features used. Such approach requires a camera which records hands movements as the PIN is typed, so unfortunately this raises safety issues.

### 3.14 Modern Trends and Commercial Applications for Keystroke Dynamics

There are several home and commercial software products that use keystroke dynamics to authenticate a user.

#### 3.14.1 Errors Made by Users and Their Correction Methods

There is a marginal tendency to recognize errors made by users and their correction methods (backspace or delete key). Many sources point this feature as a characteristic for every person, but there are not many works that confirm such a relation. Apparently the features; however useful, are difficult to implement efficiently. In [10], the users' preference of employing the left or the right shift key was tested leading to a slight improvement in the accuracy. Hence, adding new characteristic features can significantly improve the accuracy of the keystroke dynamics algorithms. The authors consider correctly typed samples much more valuable than samples with errors. Anomalies cannot provide as much information, as correct information.

### 3.14.2 Pressure-Sensitive Keyboards

Researches started looking for new features that would improve the accuracy. Naturally, using more potential hardware may introduce additional features for analysis. One of the ideas is to use pressure-sensitive keyboards. Microsoft is working on the hardware [46] and a student team contest was organized using the prototypes, searching for new ideas [47]. It is shown that pressure is even a more important characteristic than the dynamics itself. In [46], the authors constructed their own keyboard and used pressure as an additional feature, which turned out to be very helpful for the user authentication. This should not surprise anyone since it is known that online signature recognition is generally more reliable than offline. All these results suggest that the use of pressure information would greatly help in user identification.

### 3.14.3 Mobile Phone Keyboards

Some research has been done using mobile phone keyboards as input devices [49–51]. The motivation behind this is the rising popularity of mobile phones and the fact that many users do not even use a PIN to protect their device. The proposed solution to this problem is to use keystroke dynamics to authenticate users as they type text messages and phone numbers. For standard 9-key keyboard, both numerical and alphabetical inputs have been tested and the error rates reported are about 12.8% for 11-digit phone number [49] and 13% using fixed 10-character long alphabetical input [50]. Interestingly, for mobile version of QWERTY keyboard, dwell time for each key did not prove to be a reliable feature and only latency between keys was used [51]. Results were similar as for 9-key keyboard and the error rate was 12.2%.

### 3.14.4 ATM Hardware

ATM hardware was also considered [52], but rather than keystroke dynamics, keystroke motion and hand shape at different time points were analyzed. The results proved to be very good, with error rate achieved as low as 1.1%–2.7%, depending on the PIN and exact features used. This approach requires a camera which records hands movements as the PIN is typed. This is unfortunately assumed to raise safety issues.

### 3.14.5 Random Numbers Generation

Because keystroke events are triggered by human, they are—up to a certain degree—independent from external processes, and therefore are frequently used as a source of hardware-generated random numbers.

### 3.14.6 Timing Attacks on Secure Communications

Assuming that the user's keystroke dynamics characteristics are known, it may be possible to record actual keystrokes (without knowledge what actual keys are pressed) and guess what key are being pressed. Such "eavesdropping" may be done in various ways, the most simple is the recordings of typing sound, that may be also analyzed regarding the keys locations on keyboard. Another example might be a communication protocol that sends every user-typed letter separately, with well-known SSH protocol having such characteristics. It was shown in [21], that such attack may be used with moderate success and therefore pose a serious threat to seemingly secure (despite possibly strong cryptographic algorithms) communication of similar characteristics. A prevention for such attacks would be the introduction of random delay in packet transmission or introduction of "dummy" packets.

### 3.14.7 Examples of Commercial Applications

There are many home software and commercial software products that use keystroke dynamics to authenticate a user. In this section a selection of commercial applications is given, with emphasis on the used keystroke dynamics configuration.

"TypeWATCH" is an e-Biometrics solution and a patented commercial system that "continuously monitors for identity data theft attempts, by analyzing free text typing patterns of each user" [53]. The system claims to not register what is being typed, basing only on statistical analysis of dwell, flight, and typing rhythm. The system may therefore be classified as "non-fixed text with no regard to key" with continuous authentication.

"KeyTrac" [54] offers a solution for biometric password hardening ("fixed text for each user," keystrokes registered during password typing) and "non-fixed text" user identification. Biometric password hardening is supposed to be used in online stores, payments, social networks, corporate LANs and for protection of accounts. Authentication based on non-fixed text is proposed for online exams, duplicate account determination and age verification. The solution is claimed to be language independent, flexible and reliable starting from 120 keystrokes.

"Authenware®" [55] offers integrated security profile that stores and provides login data to most websites. Online education site "Coursera" [56] provides massive open online courses with implemented keystroke recognition in order to verify enrolled students identity.

### 3.15 Legal Issues

Use of key-logging software is a must for obtaining keystroke dynamics data. While some of the scientific applications may be satisfied by volunteers, a large scale tests are difficult as the use of such software may be in violation of local laws (e.g., the US Patriot Act, under which such use may constitute wire-tapping). Therefore legal advice should be obtained before attempting to use or even experiment with software and keystroke dynamic analysis, if consent is not clearly obtained from the people at the keyboard.

There are many patents in this area. Examples of several significant ones:

- Arkady G. Zilberman. Security method and apparatus employing authentication by keystroke dynamics, Patent No. 6 442 692, US B1—a device embedded in a keyboard for user identifying

- M. R. Kellas-Dicks, Y. J. Stark Keystroke dynamics authentication techniques Patent No. US 8 332 932—a large number of various metrics listed in order of diminishing discriminative power
- S. S. Bender, H. J. Postley. Key sequence rhythm recognition system and method. Patent No. 7 206 938, US Patent and Trademark Office, 2007—is based on mini-rhythms, that the user learns well and claimed to be very stable discriminators
- J. R. Young and R. W. Hammon. Method and apparatus for verifying an individual's identity. Patent No. 4 805 222, US Patent and Trademark Office, 1989—a device and method for generating storing and comparing keystroke dynamics template
- J. Garcia. Personal identification apparatus. Patent No. 4 621 334, US Patent and Trademark Office, 1986—based on a statistical comparison of individual timing vectors constructed from the time delays

## 3.16 Conclusions

Keystroke dynamics is a heavily researched topic since the beginning of the 1980s. Its use in authentication is interesting because of high acceptability and collectability. However, the efficacy is considered not sufficient enough to be used as a stand-alone identifying biometrics feature. Therefore, keystroke dynamics cannot be used for forensic purposes, as the method does not meet the European access control standards such as EN-50133-1. It specifies that FRR should be less than 1% and FAR should be no more than 0.001%. Unfortunately, the results of many researches are hardly comparable due to the use of custom or unpublished databases.

Considering keystroke dynamics permanence: an algorithm for updating the training set should be considered, as using only the initial samples would affect the system accuracy, because of generally increasing user proficiency with typing. This effect is especially striking when typing fixed or well-known text.

The approaches—regarding used information—could be classified as (1) fixed text for all users, (2) fixed text for each user, (3) non-fixed text with regard to key, and (4) non-fixed text and no regard to key. Most obvious scenario is using fixed text for authentication; however, it might be repellant to the user, if occurring frequently or with a long text. Correcting of eventual errors is usually required as it heavily affects the efficacy. Profile matching and examining is simplified in this case, as the captured and stored keystrokes might be directly comparable. A scenario regularly described in literature is using this during typing password; however it may raise security issues. Using non-fixed text is a more difficult task, however more appealing because of universality. A certain advantage of non-fixed text is possibility of using large samples transparently to the user. The most difficult scenario is when using a non-fixed text with no regard to keys. The advantage of this approach is strict secrecy, as the information what keys are being pressed should not be captured nor stored.

New keystroke dynamics features are being developed (like pressure); however, they are not widespread enough to be of practical use for the time being. Similarly to handwritten signature, when online signature is analyzed, additional features such as the pressure of the pen, its angle, and the coordinates in time can be extracted and used, with great effect [3]. Pressure or typing sound could be seen as such additional useful features. Development of mobile devices encourages keystroke dynamics for this new area, however wide use of touchscreen technologies provides new challenges.

Keystroke dynamics itself is unlikely to give satisfying results with many users; therefore, it is frequently merged with other biometric features, preferably noninvasive physiological ones in a multimodal system. One of such multifactor systems could be keystroke dynamics enhanced by a face or fingerprint recognition module. A large number of patents and industrial solutions signifies, however, great potential of this supportive behavioral biometrics.

# References

1. R. Moskovitch, C. Feher, A. Messerman, N. Kirschnick, T. Mustafic, A. Camtepe, B. Löhlein, U. Heister, S. Möller, L. Rokach, and Y. Elovici, Identity theft, computers and behavioral biometrics. In *Proceedings of the IEEE International Conference on Intelligence and Security Informatics*, 2009.

2. M. Bishop, *Computer Security*. Addison-Wesley, Boston, MA, 2002.
3. M. Adamski and K. Saeed, Online signature classification and its verification system. In *International Conference IEEE-CISIM08*, Ostrava, Czech Republic, 2008.
4. X. Li, S. Maybank, S. Yan, D. Tao, and D. Xu, Gait components and their application to gender recognition. *IEEE Transactions on Systems Man, and Cybernetics, Part C, Applications and Reviews*, 38(2), 145–155, 2008.
5. M. Obaidat and B. Sadoun, *Verification of Computer Users Using Keystroke Dynamics*.
6. S. Bleha, C. Slivinsky, and B. Hussien, Computer-access security systems using keystroke dynamics. *IEEE Transactions on Pattern Analysis and Machine Intelligence*, 12, 1217–1222, 1990.
7. O. Coltell, J. Badfa, and G. Torres, Biometric identification system based on keyboard filtering. In *IEEE International Carnahan Conference on Security Technology*, Madrid, Spain, 1999.
8. S. Haider, A. Abbas, and A. Zaidi, A multi-technique approach for user identification through keystroke dynamics. In *Systems, Man, and Cybernetics, 2000 IEEE International Conference on*, IEEE, Nashville, TN.
9. R. Joyce and G. Gupta, Identity authentication based on keystroke latencies. *Communications of the ACM*, 33(2), 168–176, 1990.
10. R. Maxion and K. Killourhy, Keystroke biometrics with number-pad input. In *Dependable Systems and Networks*, IEEE, Chicago, IL, pp. 201–210, 2010.
11. P. H. Dietz, B. Eidelson, J. Westhues, and S. Bathiche, A practical pressure sensitive computer keyboard. *Proceedings of the 22nd Annual ACM Symposium on User Interface Software and Technology*, October 4–7, ACM, New York, pp. 55–58, 2009.
12. M. Rogowski, K. Saeed, M. Rybnik, M. Tabedzki, and M. Adamski, User Authentication for Mobile Devices. In *12th IFIP TC 8 International Conference*, Springer, Kraków, Poland, 2012.
13. M. Rogowski and K. Saeed, A study on touchscreen devices: User authentication problems. In *Biometrics and Kansei Engineering*, Springer, New York, 2012.
14. J. Vacca, *Biometric Technologies and Verification Systems*, Butterworth-Heinemann, Waltham, MA, 2007.
15. M. Rybnik, M. Tabedzki, and K. Saeed, A keystroke dynamics based system for user identification. In *IEEE-CISIM08 Computer Information Systems and Industrial Management Applications*, Ostrava, Czech Republic, 2008.
16. M. Choras and P. Mroczkowski, Keystroke dynamics for biometrics identification. *Adaptive and Natural Computing Algorithm, Lecture Notes in Computer Science*, 4432, 424–431, 2007.
17. C. Loy, W. Lai, and C. Lim, Development of a pressure-based typing biometrics user authentication system, http://digital.ni.com/worldwide/singapore.nsf/web/all/9C3774162BBC5E7F862571B6000CFA1F.

18. M. Rybnik, P. Panasiuk, and K. Saeed, Advances in the keystroke dynamics: the practical impact of database quality. *Computer Information Systems and Industrial Management, Proceedings of 11th IFIP TC 8 International Conference*, Lecture Notes in Computer Science, Vol. 7564, Springer, Heidelberg, Germany, pp. 203–214, 2012.

19. P. Panasiuk and K. Saeed, Influence of database quality on the results of keystroke dynamics algorithms. *Computer Information Systems—Analysis and Technologies* (Communications in Computer and Information Science 245), Springer, Heidelberg, Germany, pp. 105–112, 2011.

20. K. Killourhy and R. Maxion, The effect of clock resolution on keystroke dynamics. In *Proceedings of the 11th International Symposium on Recent Advances in Intrusion Detection*, Springer, Heidelberg, Germany, pp. 331–350, 2008.

21. J. Ilonen, Keystroke dynamics. *Advanced Topics in Information Processing– Lecture 03–04*, Lappeenranta University of Technology, Lappeenranta, Finland, 2002.

22. Y. Wang, G-Y. Du, and F-X. Sun, A model for user authentication based on manner of keystroke and principal component analysis. In *International Conference on Machine Learning and Cybernetics*, IEEE, Dalian, China, pp. 2788–2792, 2003.

23. R. S. Gaines, W. Lisowski, S. Press, and N. Shapiro, Authentication by keystroke timing: some preliminary results. In *RAND Corporation*, Santa Monica, CA, 1980.

24. AdmitOneSecurity, [Online]. Available: http://www.admitonesecurity. com/.

25. D. Flanagan and P. Ferguson, *JavaScript: The Definitive Guide*, 5th edn. O'Reilly & Associates, Sebastopol, CA, 2003.

26. H. Wood, The use of passwords for controlling access to remote computer systems and services. In *Proceedings of the National Computer Conference*, New York, 1977.

27. M. Burnett and D. Kleiman, *Perfect Passwords*. Syngress, Rockland, MA, 2002.

28. B. Schneier, Secure passwords keep you safer, 2007. [Online]. Available: http://www.wired.com/politics/security/commentary/securitymatters/ 2007/01/72458.

29. A. K. Jain, A. Ross, and S. Prabhakar, An introduction to biometric recognition. *IEEE Transactions on Circuits and Systems for Video Technology*, 14, 4–20, 2004.

30. IBM, The next 5 in 5 innovations that will change our lives in the next five years, 2011.

31. H. P. C. and C. van Oorschot, A research agenda acknowledging the persistence of passwords. *IEEE Security & Privacy Magazine, IEEE*, 10(1), 28–36, 2012.

32. M. Rybnik, P. Panasiuk, and K. Saeed, User authentication with keystroke dynamics using fixed text. In *IEEE-ICBAKE09 International Conference on Biometrics and Kansei Engineering*, Cieszyn, Poland, 2009.

33. The OEC: Facts about the language. Available: http://www.oxforddic tionaries.com/words/the-oec-facts-about-the-language [Retrieved June 22, 2011].

34. C. Caglioti, *Preparation for an American University Program: Vocabulary Workshop*. Southampton College of Long Island University.

35. Arpeggio in World English Dictionary, [Online]. Available: http://www. dictionary.reference.com/browse/arpeggio (accessed on July 26, 2016).

36. K. Killourhy and R. Maxion, Comparing anomaly-detection algorithms for keystroke dynamics. In *Dependable Systems & Networks*, IEEE, Lisbon, Portugal, pp. 125–134, 2009.

37. J. Checco, Keystroke dynamics and corporate security, *WSTA Ticker*, Wall Street Technology Association, NJ, 2002.

38. K. Veeramachaneni, L. Osadciw, and P. Varshney, An adaptive multi-modal biometric management algorithm. *IEEE Transactions on Systems Man, and Cybernetics, Part C Applications and Reviews*, 35(3), 344–356, 2002.

39. T. Bayes and R. Price, An essay towards solving a problem in the doctrine of chance. By the late Rev. Mr. Bayes, communicated by Mr. Price, in a letter to John Canton, A. M. F. R. S. *Philosophical Transactions of the Royal Society of London*, 53, 370–418, 1762.

40. C. Loy, W. Lai, and C. Lim, Keystroke patterns classification using the ARTMAPFD neural network. *Intelligent Information Hiding and Multi-media Signal Processing*, IEEE, Kaohsiung, China, pp. 61–64, 2007.

41. J. R. Quinlan, Induction of decision trees. *Machine Learning*, 1, 81–106, 1983.

42. L. R. Rabiner and B. H. Juang, An introduction to hidden Markov models. *IEEE ASSP Magazine*, 3, 4–16, 1983.

43. S. Haykin, *Neural Networks: A Comprehensive Foundation*. Prentice Hall, Upper Saddle River, NJ.

44. M. Rybnik, M. Tabedzki, M. Adamski, and K. Saeed, An exploration of keystroke dynamics authentication using non-fixed text of various length. In *Biometrics and Kansei Engineering, 2013 International Conference on Biometrics and Kansei Engineering*, Tokyo, Japan, 2012.

45. S. Shepherd, Continuous authentication by analysis of keyboard typing characteristics. In *European Convention on Security and Detection*, Brighton, UK, 1992.

46. P. Dietz, B. Eidelson, J. Westhues, and S. Bathiche, A practical pressure sensitive computer keyboard. In *Proceedings of the 22nd Annual ACM symposium on User Interface Software and Technology*, New York, 2009.

47. UIST, Student Innovation Contest results [Online], Available: http:// www.acm.org/uist/uist2009/program/sicwinners.html, 2009.

48. H. Saevanee and P. Bhattarakosol, Authenticating user using keystroke dynamics and finger pressure. In *Consumer Communications and Networking Conference*, Las Vegas, NV, 2009.

49. N. Clarke and S. Furnell, Authenticating mobile phone users using keystroke analysis. *International Journal of Information Security*, 6(1), 1–14, 2003.

50. P. Campisi, E. Maiorana, M. Lo Bosco, and A. Neri, User authentication using keystroke dynamics for cellular phones. *IET Signal Processing,* 3(4), 333–341, 2009.

51. S. Karatzouni and N. Clarke, Keystroke analysis for thumb-based keyboards on mobile devices. In *Proceedings of the 22nd IFIP International Information Security Conference,* Sandton, South Africa, 2007.

52. A. Ogihara, H. Matsumura, and A. Shiozaki, Biometric verification using keystroke motion and key press timing for ATM user authentication. In *International Symposium on Intelligent Signal Processing and Communications,* IEEE, Tottori, Japan, 2003.

53. TypeWATCH e-biometrics security, [Online]. Available: https://www. watchfulsoftware.com.

54. KeyTrac, [Online]. Available: https://www.keytrac.net/.

55. AuthenWare, [Online]. Available: http://www.authenware.com/.

56. Coursera, [Online]. Available: https://www.coursera.org/.

# 4
# GAIT ANALYSIS

Gait analysis is one of the behavioral biometric technologies which involves people being identified purely through the analysis of the way they walk.

This chapter highlights the features, challenges, and applications of gait analysis. A brief survey is presented discussing the existing methodologies for gait analysis. This chapter also talks about some of the benchmark databases available for performance evaluation of methods for gait analysis. Finally, a partial-silhouette-based method for gait analysis is discussed along with a note on its performance in terms of computation time and memory requirements.

Biometrics is related to the science and technology of analyzing and evaluating distinguishing features from the biological traits of human beings for the purpose of verifying the identity of a person. A number of methods dealing with physiological as well as behavioural approaches to biometric verification are found in [1,2].

Apart from voice recognition [3] which has been used for human identification for years; in recent past, some new technologies such as lips tracking [4], and motion of fingers [5] have also been proposed by several researchers.

One interesting mention in this regard is Artimetrics—biometrics for artificial entities [6]. This is the field of study that identifies, classifies, and authenticates robots, software, and virtual reality agents.

Attempt has also been made in strengthening cryptographic systems with behavioral biometric [7].

Low repeatability is a major issue in biometric authentication. This requires careful preprocessing of biometric data before extracting features from it. In [8] an image denoizing technique based on Contourlet transform has been presented.

## 4.1 Human Gait Recognition

Gait recognition refers to identification of a human being analyzing the way one walks. The key to gait recognition is that every individual shows a unique spatiotemporal pattern during walking. Discriminating features can be extracted from the patterns for individual recognition. Gait analysis is one of the important behavioral biometric techniques [9,10] for identity verification.

There are three types of approaches for gait recognition: floor sensor-based, wearable sensor-based, and machine vision-based.

Floor sensor-based approaches rely on acquiring gait data with a floor mat or ground reaction force plates laid on the floor and all the subjects have to walk over for collecting data during training and also for testing.

Wearable sensor-based techniques use different motion sensors worn by the subjects for collecting and analyzing data. These techniques require expensive sensors for acquiring data.

Due to the presence of expensive sensors, these two types of gait analysis are not suitable for practical applications of human identification at a distance. However, the acquired data is much more precise compared to that of vision-based gait analysis. This is why sensor-based gait analysis has found its application in medical diagnosis.

Machine vision-based techniques [9] are the most popular methods for gait analysis by capturing the motion sequences of a human being with a camera and analyzing them for extracting discriminating features. This technique is preferred for its low-cost nature. This type of gait analysis is suitable for noncooperating subjects as "recognition at a distance" is possible without the subjects knowing the existence of the camera.

There are different approaches to machine vision-based gait analysis. "Model-based" approaches focus on static or dynamic body parameters via modeling or tracking body components such as limbs, legs, arms, and thighs. "Appearance-based" approaches focus on either shapes of silhouettes or the motion of whole human bodies, rather than modeling the whole human body or any parts of the body.

## 4.2 Features of Gait Analysis

1. Gait is a nonobtrusive technology meaning that it is unnoticeable; the subject may not be aware of the fact that gait analysis is being performed on the captured video.

2. This is a noninvasive technique. It does not require any physical contact with the subject.

3. Recognition at a distance is a unique feature of gait analysis. This makes this technology more attractive than other biometric techniques [11].

4. Being a nonobtrusive technique, gait analysis works well for noncooperating subjects.

5. Gait analysis can be performed on low-resolution videos.

## 4.3 Applications of Gait Analysis

There are several applications of gait recognition. Some of them are listed below.

1. Surveillance and security applications for human identification [12].

2. Doctors use gait analysis for detecting anomalies [13] in the way a person walks. Gait is used to identify onset of medical conditions that affect the motor capabilities of a person such as in Parkinson's disease and multiple sclerosis where muscle control is debilitating. In most of the cases, these diseases are irreversible, however early identification allows earlier treatment and therapy.

3. *Clinical applications*: Gait analysis has found its application for clinical usage. Making use of wearable sensors, patient's gait data can be collected during checkup that will help in planning of treatments. Gait analysis can also be used to assess effects of treatment. This can affect surgical decisions; decrease the cost of care by reducing the need for pre- and post-operative clinic visits and surgical interventions. Gait analysis can be used as a mean of deducing the physical well-being of people [13].

4. Gait analysis is also used in gender detection [14,15].

5. *In sports*: By placing wearable sensors on body segments, an athletes' performance and areas for improvements can be analyzed. Detection of incorrect posture will help preventing sports injuries.

## 4.4 Gait Cycle

A gait cycle is the collection of all the temporal sequences contained between two consecutive heel strikes by the same leg. One gait cycle

Mid-stance　　　　　　Double-support

**Figure 4.1**　Gait cycle.

consists of two steps: one of the right foot and one of the left foot. Each step is divided into two phases: stance phase and swing phase. Stance phase comprises almost 60% and swing phase almost 40% of a gait cycle (Figure 4.1).

Right stance phase begins with the heel contact of the right foot and ends with the toe-off of the same foot. Right swing phase begins with the toe-off of the right foot and ends with the heel contact of the same foot. Each stance phase is divided into four components: heel contact, mid-stance; active proportion, and passive proportion. End of each phase marks the beginning of the next. Right heel contact begins with the heel contact of the right foot and ends with the toe-off of the left foot. Right mid-stance begins with the toe-off of the left foot and ends with the heel lift of the right foot. Having defined a cycle, gait can be defined as follows—gait is a total walking cycle.

A stride can be defined to consist of two double-support and two mid-stance frames. Double support is the stance when both the legs are furthest apart and touching the ground. Depending upon which leg is in front this can be categorized as left and right double-support frame.

Mid-stance is the stance when legs are closest together with one swinging leg just passing the other upright planted leg. Depending upon which leg is in upright position, this stance can be categorized as left mid-stance and right mid-stance.

It is important to identify the cycle as it contains all the temporal sequences of a moving human body that need to be analyzed for discriminating features for gait analysis. After a cycle the walking sequence contains the temporal sequences of another cycle which is a repetitive pattern.

### 4.5　Describing a Stance

During walking of a person, the distribution of force (body weight) at the bottom of human feet can be subdivided as follows at different times of a stance.

1. *Initial contact (heel strike)*: The majority of the body weight is on the left foot. With heel contact of the right foot, the weight of the whole body is slightly transferred from left to right foot.
2. *Mid-stance (foot flat)*: This is the mid-stance when the entire sole touches the ground. The body weight is exerted all over the sole. This is the time when the body weight of a walking human being is transferred from one leg to another.
3. *Terminal stance*: The body weight is shifted toward the front of the sole.
4. *Toe off*: This is the pre-swing phase. The other foot (left in this case) touches the ground and the partial weight of the body is exerted at the toes of one foot (right in this case) and the other foot (left in this case) balances the body weight.

It is interesting to note that the distribution of force (body weight) over time by the same person in two different conditions (e.g., with a shoe and bare footed) show different nature. This is the reason why gait analysis is a challenging task under different shoe conditions.

## 4.6  Why Does Gait Change from Person to Person or from Time to Time?

Behavioral biometric attributes sometimes depend on the physical aspects of a person. The body parts primarily accounted for the human locomotion are the legs. The features that seem unique to a person are the following: joint angle between the upper and lower legs, relationship between the knee joints and the feet over time, elevation of knee joint over the ankle (i.e., vertical distance between knee and ankle) shows a distinctive temporal pattern, and transition from swing leg to stance leg is noticeably different across different people over time.

## 4.7  A Brief Review of the Literature on Human Gait Recognition

This section briefly represents a review of existing literature on gait recognition. State-of-the-art methods dealing with different challenges of gait analysis, for example, changes in appearance [16], speed variance [17], clothing variance [18], and some multimodal techniques [19,20] along with some other recent methods are also discussed at the end of this section.

A survey of different techniques for vision-based human identification can be found in [9]. Computer vision-based gait analysis can be broadly divided into two categories: model-based approaches and appearance-based approaches.

Model-based approaches obtain a series of static or dynamic body parameters via modeling or tracking body components such as limbs, legs, arms, and thighs. Gait signatures derived from these model parameters are employed for identification and recognition of an individual.

In [21], a two-step, model-based approach to gait recognition is presented which employs a five-link biped locomotion human model. The extraction of gait features from image sequences is performed using the Metropolis–Hasting method. Hidden Markov models (HMMs) are then trained based on the frequencies of the feature trajectories from which recognition is performed. Primary model-based approaches employ static structure parameters of body as recognition features.

BenAbdelkader et al. [22] presented structural stride parameters consisting of stride and cadence. The cadence is estimated via the walking periodicity, and the stride length is calculated by the relation of traveled distance and walking steps.

Bobick and Johnson [23] calculated four distances of human bodies, namely the distance between the head and foot, the distance between the head and pelvis, the distance between the foot and pelvis, and the distance between the left foot and the right foot. They used the four distances to form two groups of static body parameters and reveal that the second set of parameters are more view-invariant compared to the first set of body parameters.

Model-based approaches are mostly view-invariant and scale-independent. Their advantages are significant for practical applications, because it is unlikely that reference sequences and test sequences are taken from the same viewpoint. However, model-based approaches are sensitive to the quality of gait sequences. Thus, gait image sequences of high quality are required to achieve a high accuracy. Another disadvantage of the model-based approach is its large computation and relatively high time costs due to parameters calculations. These approaches have not reported high performance on common databases, partly due to the self-occlusion caused by legs and arms crossing [16].

Appearance-based approaches focus on either shapes of silhouette or the whole motion of human bodies, rather than modeling the whole human body or any parts of body.

Boulgouris et al. identified gait analysis as one of the challenging tasks [24] for biometric identification. As an initial attempt to solve the problem, a baseline algorithm is proposed by Sarkar et al. [25]. The baseline method uses the silhouettes of human body themselves as features, which are scaled and aligned before extracting features.

Han and Bhanu [26] present the idea of the gait energy image (GEI) for individual recognition. GEI converts the spatiotemporal information during one walking cycle into a single 2D gait template, which avoids matching features in temporal sequences. Paper [26] addresses the problem of lack of training templates by combining statistical gait features from real and synthetic templates. GEI is comparatively robust to noise by averaging images of a gait cycle. However, it loses the dynamical variation between successive frames. Jianyi and Nanning [27] develop the gait history image (GHI) to retain temporal information as well as spatial information.

Chen et al. [28] proposed the frame difference energy image (FDEI) based on GEI and GHI to address the problem of silhouette incompleteness. They calculated the positive portion of frame difference as positive values of the subtraction between the current frame and the previous frame. FDEI is defined as the sum of GEI and the positive portion.

Wang et al. [12] transformed the 2D contour of silhouette to a 1D signal using the distance between pixels along the contour and the shape of the centroid. However, these 1D signals are easily affected by the quality of silhouettes.

The appearance-based model-free approaches have the advantage of low computational costs compared to model-based approaches. However, these are often sensitive to the quality of silhouettes. Besides, these approaches are usually not robust to viewpoints and scale.

Although significant recognition accuracy has been achieved in human identification using gait analysis, researchers also explore the possibility of using both face and gait in enhancing human recognition at a distance in outdoor conditions [11]. Individual performance of gait- and face-based biometrics at a distance under outdoor illumination conditions, walking surface changes, and time variations are poor; however it has been shown that recognition performance is significantly enhanced by combination of face and gait [11].

There are several factors affecting the performances of detection algorithms for computer vision-based gait recognition systems. Paper [29] describes a large set of video sequences acquired for investigating important dimensions of the problem, such as variations due to viewpoint, footwear, and walking surface. Results obtained in [29] suggest that differences in footwear or walking surface type between the gallery and probe video sequence affect performance significantly.

In [30] a simple method for human identification based on body shape and gait is presented. This baseline method provides a lower bound against which to evaluate more complicated procedures. A viewpoint-dependent technique based on template matching of body silhouettes is presented. Cyclic gait analysis is performed to extract key frames from a test sequence. These frames are compared to training frames using normalized correlation, and subject classification is performed by nearest-neighbor matching using correlation scores. The approach implicitly captures biometric shape cues such as body height, width, and body part proportions, as well as gait cues such as stride length and amount of arm swing. The method is verified on four databases with varying viewing angles, background conditions (indoors and outdoors), and walking styles.

In [31], a representation of gait appearance is introduced for the purpose of person identification and classification. This gait representation is based on extracting moment features from human silhouettes in video taken from an orthogonal view. Despite its simplicity, the resulting feature vector contains enough information to perform well on human identification and gender classification tasks. In addition to human recognition under various conditions, [31] the gait works well for gender classification based on the appearance features using a support vector machine (SVM).

In [32], a method is presented for describing gait contour by using Fourier descriptors. The Fourier descriptors are used to make a periodical analysis on the height and width ratio of the gait image and to solve the problems resulting from an image sequence of different gait cycles by using dynamic time warping. Silhouette-based gait recognition is one of the most popular methods for recognizing moving shapes.

Veres et al. [33] have aimed at investigating the important features in silhouette-based gait recognition from the point of view of statistical analysis. It is shown that the average silhouette includes a static component of gait (head and body) as one of the most important image part. The dynamic component of gait (swings of legs and arms) is also considered as an important information for gait analysis. It implies that the static component of gait can be used for human recognition while suggesting that ignoring the dynamic part of gait can result in loss of recognition accuracy in some cases. Consequently, the importance of better motion estimation is underlined.

Researchers have addressed the issue of gait recognition under view invariance. A framework is presented in [34] for this task with training gait sequences from multiple views. The most important problem in the framework is about the optimal choice for the training views, that is, how many views are enough to ensure a satisfying overall performance and how to combine these views to achieve the optimal performance. The use of mean of radon transforms of the silhouettes is described as the descriptor which is very competent for view-invariant applications. Moreover, the combination of class correlation and view correlation is applied to score level fusion of results from different views. The CASIA database is used for experiments.

A multimodal biometric recognition system is presented in [35] using three modalities including face, ear, and gait. The system is based on Gabor and principal component analysis (PCA) feature extraction methods with fusion at matching score level. The performance of the approach has been studied under three different normalization methods (min-max, median-MAD, and z-score) and two different fusion methods (weighted sum and weighted product). According to the experimental results [35], the method exhibits good recognition performance and outperforms unimodal systems.

A robust algorithm is presented in [36] for human identification from a sequence of segmented noisy silhouettes in a low-resolution video. The algorithm enables automatic human recognition from model-based gait cycle extraction based on the prediction-based hierarchical active shape model (ASM). Instead of directly analyzing the gait pattern, a set of relative model parameters is extracted. The feature extraction function consists of motion detection, object region detection, and ASM, which alleviate problems in the baseline algorithm such as background generation, shadow removal, and higher recognition rate. Performance of the algorithm has been evaluated by using the HumanID Gait Challenge dataset with different realistic parameters including variations in viewpoints, shoes, different surfaces, and carrying conditions and time.

Multimodal [37,38], multiview, and multistance [39], subgait characteristics-based [40] gait recognition systems also exist. A methodology for gait recognition using ranking is presented in [41].

A novel approach for gait recognition based on the matching of body components is presented in [42]. The human body components are studied separately and are shown to have unequal discrimination power. Several approaches are presented for the combination of the results obtained from different body components into a common distance metric for the evaluation of similarity between gait sequences. A method is also proposed [42] for the determination of the weighting of the various body components based on their contribution to recognition performance.

Feature subset selection is an important preprocessing step for pattern recognition, to discard irrelevant and redundant information, as well as to identify the most important attributes. An attempt has been made [43] toward investigating a computationally efficient solution to select the most important features for gait recognition. The technique is based on mutual information, which evaluates the statistical dependence between two random variables and has established relation with the Bayes classification error. Sequential selection method based on mutual information provided an effective solution for high-dimensional human gait data. The experiments are carried out based on a 73-dimensional model-based gait features set and on a 64 by 64 pixels model-free gait symmetry map on the Southampton HiD Gait database. Results show that the method outperforms methods based on correlation and analysis of variance.

Another approach to gait recognition is presented by Murat and Murat [44]. Binarized silhouette of a motion object is first represented by four 1D signals that are the basic image features called the distance vectors. The distance vectors are differences between the bounding box and the contour of the silhouette. Vectors are generated using four projections to silhouette. Fourier transform is employed as a preprocessing step to achieve translation invariance for the gait patterns accumulated from silhouette sequences that are extracted from the walking sequence of the subjects under different speed and/or different time. Then, eigenspace transformation is applied to reduce the dimensionality of the input feature space. SVM-based pattern classification technique is then used on the lower-dimensional eigenspace for recognition. The input feature space is alternatively constructed by using two different approaches. The four projections (1D signals) are independently classified in the first approach. A fusion task is then applied to produce the final decision. In the second approach, the four projections are concatenated to have one vector and then pattern classification with one vector is performed in the lower-dimensional eigenspace for recognition.

Variations in clothing alter the appearance of an individual making the problem of gait identification much more difficult. If the type of clothing differs between the gallery and the probe, certain parts of the silhouettes are likely to change and the ability to discriminate subjects decreases with respect to these parts. A part-based approach, therefore, has the potential of selecting the appropriate parts. In [45], a method is described for part-based gait identification in the light of substantial clothing variations. The whole human body is divided into eight sections, including four overlapping ones, since the larger parts have a higher discrimination capability, while the smaller parts are more likely to be unaffected by clothing variations. Furthermore, as there are certain clothes that are common to different parts therefore a categorization is made for items of clothing that groups similar clothes. Afterward the discrimination capability is used as a matching weight for each part and the weights are controlled adaptively based on the distribution of distances between the probe and all the galleries. Finally, the method is tested with large-scale gait dataset with clothing variations.

Recent gait recognition methods mostly focus on various aspects of gait including making use of multimodal systems. Paper [19] provides a means for multimodal gait recognition, by introducing the freely available TUM Gait from Audio, Image and Depth (GAID) database. This database simultaneously contains RGB video, depth, and audio. With 305 people in three variations, it is one of the largest to date. To further investigate challenges of time variation, a subset of 32 people is recorded a second time. A standardized experimental setup is defined for both person identification and for the assessment of the soft biometrics age, gender, height, and shoe type. For all defined experiments, several base-line results on all available modalities are presented. These effectively demonstrate multimodal fusion being beneficial to gait recognition.

The advantage of gait over other biometrics such as face or finger-print is that it can operate from a distance and without cooperation of the subjects. However, this also makes gait subject to changes in various covariate conditions including carrying, clothing, surface, and view angle. Existing approaches attempt to address these condition changes by feature selection, feature transformation, or discriminant subspace learning. However, they suffer from lack of training samples from each subject, can only cope with changes in a subset of conditions with limited success, and are based on the invalid assumption that the covariate conditions are known a priori. They are thus unable to perform gait recognition under a genuine uncooperative setting.

The identification of a person from gait images is generally sensitive to appearance changes, such as variations of clothes and belongings. One possibility to deal with this problem is to collect possible appearance changes of the subjects in a database. However, it is almost impossible to predict all appearance changes in advance. In [16], a method is presented which allows robustly identifying people in spite of changes in appearance, without using a database of predicted appearance changes. First, the human body image is divided into multiple areas, and features for each area are extracted. Next, a matching weight for each area is estimated based on the similarity between the extracted features and those in the database for standard clothes. Finally, the subject is identified by weighted integration of similarities in all areas.

In [20], a novel human-identification scheme from long range gait profiles in surveillance videos is presented. The role of multiview gait images acquired from multiple cameras and the importance of infrared and visible range images in ascertaining identity is investigated along with the impact of multimodal fusion, efficient subspace features, and classifier methods. The role of walking style in enhancing the accuracy and robustness of the identification systems is also investigated.

In [41], a new approach is presented which casts gait recognition as a bipartite ranking problem and leverages training samples from different classes/people and even from different datasets. This makes the approach suitable for recognition under a genuine uncooperative setting and robust against any covariate types, as demonstrated by the experiments in [41].

Robust gait recognition is a challenging problem, due to the large intra-subject variations and small inter-subject variations. Out of the covariate factors like shoe type, and carrying condition, elapsed time, it has been demonstrated that clothing is the most challenging covariate factor for appearance-based gait recognition. For example, long coat may cover a significant amount of gait features and make it difficult for individual recognition.

Paper [18] proposes a framework using random subspace method (RSM) for clothing-invariant gait recognition by combining multiple inductive biases for classification. Even for small size training set, this method can achieve promising performance. Experiments are conducted on the OU-ISIR Treadmill dataset B which includes 32 combinations of clothing types, and the average recognition accuracy is reported to be 80%, which indicates the effectiveness of the method.

Paper [45] describes a large-scale gait database comprising the Treadmill Dataset. The dataset focuses on variations in walking conditions and include 200 subjects with 25 views, 34 subjects with 9 speed variations from 2 to 10 km/h with 1 km/h interval, and 68 subjects with at most 32 clothes variations. The range of variations in these three factors is significantly larger than that of previous gait databases, and therefore the Treadmill Dataset can be used in research on speed-invariant gait recognition. Moreover, the dataset contains more diverse gender and ages than the existing databases and hence it enables evaluating gait-based gender and age group classification in more statistically reliable way.

View transformation in gait analysis has attracted more and more attentions recently. However, most of the existing methods are based on the entire gait dynamics, such as GEI and the distinctive characteristics of different walking phases are neglected. Hu et al. [39] proposes a multiview multistance gait identification method using unified multiview population HMMs (pHMM-s), in which all the models share the same transition probabilities. Hence, the gait dynamics in each view can be normalized into fixed-length stances by Viterbi decoding. To optimize the view-independent and stance-independent identity vector, a multilinear projection model is learned from tensor decomposition. The advantage of using tensor is that different types of information are integrated in the final optimal solution. Extensive experiments show that the algorithm achieves promising performances of multiview gait identification even with incomplete gait cycles.

Gait recognition algorithms often perform poorly because of low-resolution video sequences, subjective human motion, and challenging outdoor scenarios. Despite these challenges, gait recognition research is gaining momentum due to increasing demand and more possibilities for deployment by the surveillance industry. Therefore, every research contribution which significantly improves this new biometric is a milestone. Paper [40] proposed a probabilistic subgait interpretation model to recognize gaits. A subgait is defined as part of the silhouette of a moving body. Binary silhouettes of gait video sequences form the basic input of this approach. A novel modular training scheme has been introduced to efficiently learn subtle subgait characteristics from the gait domain. For a given gait sequence, useful information is obtained from the subgaits by identifying and exploiting intrinsic relationships using Bayesian networks. By incorporating efficient inference strategies, robust decisions are made for recognizing gaits.

Multimodal biometric has the ability to improve the performance in biometric recognition systems. Paper [37] presents multimodal biometrics using feature level fusion strategy. Walking person's face and gait data are captured and represented as combined feature vector. PCA is used to reduce the dimensionality of the feature vectors. Experiments show that when only gait feature of individual are used, recognition accuracy of only 67% is achieved. However, when combined gait and face features are considered the performance is improved up to 90%.

The research trend in multimodal biometric systems highlights the limitations in behavioral biometric approaches like gait analysis—relative low recognition accuracy due to low repeatability. This suggests the usefulness of a low-cost gait recognition methodology with considerable high recognition accuracy over a complex, resource-intensive method with high accuracy level.

## 4.8 Research Challenges

There are many factors both external and internal that pose challenges to real-life gait analysis.

### 4.8.1 External Factors

Factors which are not related with a human being but on something else are external factors.

Viewing angles of the camera (e.g., frontal view and side view) has an impact on gait analysis. Normally, lateral or side view has proven to be the best orientation of camera for human gait analysis. Some methods are view invariant.

Accuracy of a gait recognition system depends on ambient lighting conditions (e.g., day/night) in which the motion sequence of the subject is captured. Lighting condition has a strong effect on the object segmentation phase. A poorly segmented human body will generate inaccurate silhouette resulting in poor recognition accuracy.

Consideration of the environment, for example, outdoor/indoor environments (e.g., sunny/rainy days) and so on, is important for gait analysis. Real-life situation is always hard to handle due to presence of different kinds of noises.

Gait analysis becomes very difficult under different clothing conditions (e.g., skirts/trousers/gown) of the subjects. The reason is that some gait features may not be extracted for two different wearing conditions. As for example, if a subject walks with trousers, the leg dynamics can be easily extracted from the temporal sequences of the silhouettes. On the other hand, when the same subject is walking wearing a long gown, the legs of the subject are occluded by the cloth imposing difficulty in gait recognition.

Gait also changes depending on the condition of the surface one walks over (e.g., hard/soft, dry/wet grass/concrete, level/stairs, etc.). Walking style also gets affected by the different types of shoes one puts on (e.g., mountain boots and sandals).

Another important issue is the carrying conditions. Gait of a person changes depending on different object carrying conditions, for example, backpack, briefcase, and so on.

Real-time object segmentation is very tough due to the presence of various types of noises in the environment (e.g., illumination change in the scene and subsequent failure in detecting the moving human body). Occlusion is another important issue to consider during gait analysis. A person can be occluded by other objects/persons present in the scene. The features cannot be extracted from an occluded body part of a person leading to misclassification and/or degradation of recognition accuracy.

### 4.8.2 *Internal Factors*

Such factors cause changes of the natural gait due to sickness (e.g., foot injury, lower limb disorder/disease, etc.) or other physiological changes in body due to aging, drunkenness, pregnancy, gaining or losing weight, and so on.

## 4.9 Gait Databases for Research

A number of databases are provided by a number of institutions for research on gait analysis.

### 4.9.1 *CASIA-A*

Dataset provided by the Chinese Academy of Sciences Institute of Automation includes silhouettes of 20 subjects and 4 sequences for each viewing angle per subject: 2 sequences for one direction of walking, the other 2 sequences for the reverse direction of walking. Frames: 180 × 220.

All those gait sequences were captured twice on two different days, in an outdoor environment. All subjects walk along a straight-line path at free cadences in three different views with respect to the image plane (Figure 4.2).

### 4.9.2 *CASIA-B*

It is a large multiview dataset containing 124 subjects from 11 views. Subjects walked in four different conditions: normal walking, slow walking, fast walking, and normal walking with a bag (Figure 4.3).

| Lateral view | Oblique view | Frontal view |

**Figure 4.2**   CASIA-A dataset, three views. (From L. Wang et al., *IEEE Trans. Pattern Anal. Machine Intell.*, 25, 1505–1518, 2003. With permission.)

**Figure 4.3** CASIA-B dataset, eleven views. (Courtesy of Center for Biometrics and Security Research, Beijing, China, http://www.cbsr.ia.ac.cn/english/GaitDatabases.asp.)

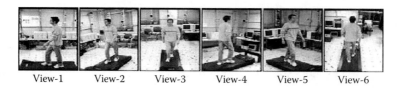

View-1    View-2    View-3    View-4    View-5    View-6

**Figure 4.4** CMU MoBo different view angles. (From R. Gross and J. Shi, The CMU Motion of Body [MoBo] database, Tech. Report CMU-RI-TR-01-18, Robotics Institute, Carnegie Mellon University, 2001. With permission.)

### 4.9.3 CMU MoBo

This dataset is provided by Carnegie Mellon University, Institute of Robotics. It includes 25 subjects (23 males, 2 females) walking on a treadmill. Each subject is recorded performing four different types of walking: slow walk, fast walk, inclined walk, and slow walk holding a ball. There are about 8 cycles in each sequence, and each sequence is recorded at 30 frames (486 × 640) per sec. It contains six simultaneous motion sequences of the 25 subjects from six different viewing angles (Figure 4.4).

### 4.9.4 USF Dataset

This database is provided by the University of South Florida. It contains 1870 sequences on 122 subjects on two different shoe types. It also considers different carrying conditions like with/without a briefcase;

different surface conditions grass/concrete surface; different camera views, for example, left/right viewpoint. Video is captured in two different time instants. It contains sequences taken in outdoor environment.

### 4.9.5  Southampton Dataset

"Southampton Human ID at a Distance" database is the contribution by the University of Southampton. It consists of two major segments—small and large databases. Small database consists of 12 subjects walking around an insidetrack at different speeds, wearing different shoes, clothes, and without or with various bags. Large database were people filmed walking outside, inside track and inside treadmill from six different views on 114 subjects with a collection of more than 5000 sequences.

### 4.9.6  3D Dataset

This is a novel 3D gait dataset. Samples are captured when subjects walked through a multibiometric tunnel.

### 4.9.7  UMD Dataset

This dataset is provided by University of Maryland. It includes 55 subjects and 122 motion sequences are taken from four different view points in an outdoor environment.

### 4.9.8  TUM-IITKGP Dataset

This dataset is provided by the Technical University of Munich, Germany, and Indian Institute of Technology Kharagpur, India. It contains sequences with regular walk, walking subjects putting their hands in pocket to inhibit the motions of hands, walking with a backpack, walking wearing a gown. This is the only public dataset to consider samples taken considering dynamic and static occlusions. This dataset contains 35 subjects and 840 sequences. The environment is indoor and hallway.

### 4.9.9  OU-ISIR Database

This is a large-scale gait database comprising the treadmill dataset. The dataset focuses on variations in walking conditions and include

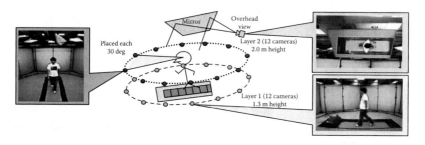

**Figure 4.5** Multiview synchronous gait capturing system (OU-ISIR). (From Y. Guan et al., Robust clothing-invariant gait recognition, *International Conference on Intelligent Information Hiding and Multimedia Signal Processing*, IEEE, Piraeus, Greece, pp. 321–324, 2012. With permission.)

200 subjects with 25 views, 34 subjects with 9 speed variations from 2 to 10 km/h with 1 km/h interval, and 68 subjects with at most 32 clothes variations. The range of variations in these three factors is significantly larger than that is available in previously presented gait databases, and therefore the treadmill dataset can be used in research on invariant gait recognition. Moreover, the dataset contains more diverse gender and ages than the existing databases and hence it enables to evaluate gait-based gender and age group classification in more statistically reliable way (Figure 4.5).

A total of 25 cameras; 12 encircled cameras at layer 1, 30 degree apart placed at a 1.3 m height from the ground; other 12 at layer 2 encircled with similar angular separation 2 m height and one overhead camera.

## 4.10 Gait Recognition Using Partial Silhouette-Based Approach

Silhouette-based gait analysis is a well-established biometric approach for human identification. Over the years, researchers have proposed a number of gait recognition approaches based on the entire silhouette of human body. These approaches are proven to give good recognition accuracies. However, the generation of feature vectors and subsequent classification depend on information extracted from the entire object silhouette involving handling of considerably large amount of data. This chapter discusses about an innovative partial-silhouette-based method [48] for gait analysis considering the fact that the partial silhouette of a human body often contains sufficient discriminating information for gait analysis. This method is based on extracting features from the portion/part of the human silhouette that contains one of the most dynamic features of gait—the swinging hands of a person. This method has been experimentally verified using two standard widely used public gait datasets.

Most of the vision-based techniques that involve analysis of gait based on human silhouette are based on representing the boundary of the silhouette with a boundary descriptor for feature extraction. Most of the techniques involve boundary warping [12], Fourier descriptors [32], or projection-based method [44] for representing the boundary of the silhouette. These methods require either representing the entire contour of the object [12] or a significant portion of it [32,44]. However, it has been observed that the information represented by the whole boundary is sometimes redundant for gait analysis. Some sections of the moving silhouette carry information that is much more useful for recognition than the other ones. These parts of human body (e.g., region of legs and hand swing) contain most important dynamic features for gait analysis [16,33].

### 4.10.1 Motivation of the Partial Silhouette-Based Approach

There are many factors affecting the recognition accuracy for gait analysis, namely the viewing angle of the camera (e.g., lateral view, frontal view, oblique view, etc.) [34], different clothing conditions of the subjects (normal outfit, wearing a coat or gown, etc.) [16,18], walking speed of the subjects (fast vs. slow walk) [17], different carrying conditions (holding a bag/briefcase or walking with free hands) [24,25], and so on.

The accuracies of the existing methodologies largely depend on the view of the camera. The method which is suitable for lateral view may produce poor results for frontal or oblique views. Security systems that demand high accuracy level may not rely only on gait analysis but on multimodal biometric technologies.

Apparently, these highlight the limitations of gait analysis as an acceptable biometric technology for systems requiring very high recognition accuracies. However, relatively less repeatability is a common problem not only for gait analysis but also for other behavioral biometric techniques (e.g., keystroke dynamics). This leads to systems making use of multimodal systems where more than one biometric technology are integrated together to ensure a very high accuracy level of individual recognition. Multimodal systems may involve many combinations of gait and other biometric modalities. Some of the known researches are as follows: gait and face recognition [11,38], gait along with face and ear biometrics [35], and gait and other traits [19]. Trends also show that researchers are involved in multimodal images (e.g., normal and infrared imaging) to improve the accuracy of gait recognition [38]. However, the complexity of a multimodal system will be very high if individual biometrics (e.g., gait, face, ear, etc.) are resource-intensive, making lots of computations and requiring a large amount of storage for processing. A low-cost, yet relatively accurate method for gait recognition could be useful in this scenario.

In view of this, a novel methodology for gait recognition is presented in [48]. This is based on partial silhouette of a human body and is essentially an appearance-based approach that aims at identifying a person from silhouette analysis considering the portion of body containing the dynamics of swinging hands. The technique presented here is insensitive to translation by virtue of the silhouette boundary descriptor. Rotation invariance can be achieved by selecting the same starting point for calculating the signature. Scale invariance is obtained by using normalized distance. Moreover, as the new method considers a portion of the object silhouette, it works well on incomplete or noisy and distorted silhouettes as long as the section containing the hand dynamics could be extracted correctly.

### 4.10.2 Dynamic Features of Gait—Why Partial Silhouette?

Dynamic component of gait is typically contained in the region of the hands and legs of a moving human body. The justification of choosing the partial silhouette for gait recognition is described below with the help of two image descriptors namely GEI [43] and width vector [10].

*GEI*: Given the preprocessed binary gait silhouette images $B_t(x, y)$ at time $t$ in a sequence, the gray-level GEI is defined as follows:

$$G(x, y) = \frac{1}{N} \sum_{t=1}^{N} B_t(x, y)$$

Here, $N$ is the number of frames in a complete cycle of a silhouette sequence, $t$ is the frame number in the sequence (moment of time), and $x$ and $y$ are values in the 2D image coordinate. Figure 4.6 shows a sequence of silhouette images in a gait cycle of a subject. The right most image is the corresponding GEI.

In the GEI image, the regions of the hands and legs show shades of gray representing the portion of the silhouette that contain the dynamic component of gait. GEI implies that during walking, the portion of hands and legs captures the maximum swing compared to other body parts.

*Width Vector*: The width of the outer contour of the binary silhouettes of a human being can be considered as the feature vector. Width vectors retain the physical structure of the subjects as well as the swing of the limbs and other details of a subject. The width along a given row (scan line) of an image is computed as the difference in the locations of the right-most and the left-most boundary pixels in that row and a width vector is generated for each frame. The variation of each component of the width vector can be regarded as the gait signature of that subject. A temporal width vector plot is shown in Figure 4.7. The width vectors are plotted along the $y$-axis, and the $x$-axis contains the temporal sequences of the silhouettes. It is clear from the figure that the width vector is roughly periodic and gives the extent of movement of different parts of the body. The brighter a pixel, the larger is the value of the width vector in that position implying that the maximum swing in that body component is accumulated at the portions of hands and legs of a human body during walking (Figure 4.7).

The above discussion implies that the dynamic features of gait can be captured from the portions of the human silhouettes containing the swings of hands.

**Figure 4.6** Sequence of silhouette images in a gait cycle of a subject and the right most image is the corresponding GEI.

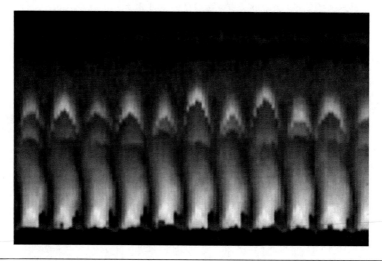

**Figure 4.7**  Temporal plot of the width vectors. (From A. Kale et al., *IEEE Trans. Image Process.*, 13(9), 1163–1173, 2004. With permission.)

*4.10.3 Partial Silhouette-Based Methodology*

The objective of this work is to present a low-cost method for human gait recognition with a substantial accuracy level. However, if a system demands higher accuracy, then the present methodology is the ideal method to be integrated with other modalities like face recognition or ear biometrics due to its low computational complexity.

Gait features from silhouettes can be separated into static appearance features and dynamic gait features, which reflect the shape of human body and the way how people move during walking, respectively. The region of the swinging hands of a human body retains dynamic features containing sufficient discriminating characteristics for human identification using gait analysis. The partial silhouette-based method is a part-based approach for gait recognition based on extracting the part of the swinging hands of human body from the whole human silhouette.

*4.10.4 Preprocessing for Removing Noise*

As shown in Figure 4.8a, the binarized frames contain noises (encircled at the top left) along with the object of interest (encircled at the bottom right). Before extracting the landmark frames, noises are to

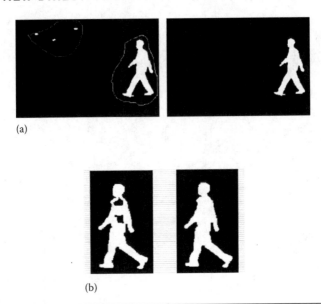

**Figure 4.8**   Noise removal, using (a) morphological processing and (b) boundary filling.

be removed. Morphological processing is performed to remove this noise. It is seen that often the objects of interest contain noises which are removed by applying boundary fill methods.

*4.10.5 Gait Cycle Detection and Extraction of Landmark Frames*

Gait consists of a total walking cycle (Figure 4.9a). A gait cycle contains all the frames in the sequence of two consecutive heal strikes between the same leg. In a cycle there are four landmark (key) frames known as double-supports and mid-stances. In a double-support

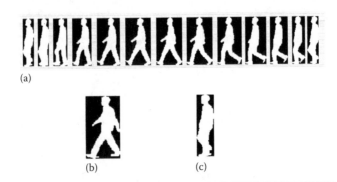

**Figure 4.9**   (a) Gait cycle. Landmark frames (b) double-support and (c) mid-stance.

frame both legs are furthest apart and touching the ground. It is classified as left and right double-support depending on which leg is in front. In a mid-stance frame, legs are closest together with one swinging leg just passing the other upright, planted leg. Depending on which leg is planted, mid-stance is also classified as left and right mid-stance frames.

Walking consists of a sequence of frames contained in a gait cycle. Detection of a cycle is important. A cycle contains the information that is required for feature extraction. Another cycle is just the repetition of the temporal sequences of human silhouettes. The methodology presented in [30] is used for detecting a cycle and identifying the landmark frames. The height and width ratio of the bounding box of a moving human body is used as signal to identify the cycle and double-support and mid-stance frames from the sequence of silhouettes of a moving object (human being). The width of the bounding box is used as a signal to measure the distance between two legs. The width of the bounding box is the maximum for the double-support frames and minimum for the mid-stance frames (shown in Figure 4.9b and c, respectively). Two double-support and two mid-stance frames are extracted from a gait cycle using this methodology. Extracted landmark frames are further used for feature extraction and classification.

## 4.11 Extraction of Partial Silhouette

This is a silhouette-based method. The dependency of color or texture of dresses of a moving human being is removed as a binary silhouette is considered. The binary silhouette also decreases the dimension of data to be handled.

There are different part-based methodologies in the literature [14,16,42,45]. It is found [33] that during walking the human silhouettes contain both static features (head, torso) and dynamic features (swinging hands and legs) of gait in different body parts. The partial silhouette-based methodology is a part-based approach in which only a portion of the whole human body (e.g., the portion of the swinging hands) is extracted and analyzed for gait recognition. Extracting a portion from the silhouette requires segmenting the entire silhouette image. The following steps have to be performed.

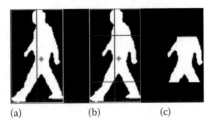

(a)      (b)      (c)

**Figure 4.10**  Image segmentation. (a) Human silhouette and the corresponding bounding box. Plus sign indicates the centroid. Vertical line is the height of the bounding box. (b) Horizontal lines are equidistant from the centroid. The portion of the silhouette contained within these two lines (horizontal segment) is to be extracted. (c) Extracted portion of the silhouette.

### 4.11.1 Bounding Box

The bounding box is computed for the whole binary silhouette. The bounding box is the minimum rectangle that most tightly contains the human silhouette. After the object is put into a bounding box, it is size-normalized to a fixed height keeping the aspect ratio of the image same. In Figure 4.10a and b the rectangular boxes shown with white lines are the bounding boxes. The height of the bounding box is calculated (the length of the vertical lines in Figure 4.10a and b). Let the height of the bounding box be *box_height*.

### 4.11.2 Image Segmentation

The entire silhouette image is to be segmented horizontally. It is required to decide the portions of the image to be extracted. The centroid of the binary silhouette is calculated. The centroid is shown in Figure 4.10a and also in Figure 4.10b. Let the value of the *y*-coordinate of the centroid be *y_cent*.

The segment height (*seg_height*) is calculated as one-fourth of the height of the bounding box as follows: *seg_height = box_height*/4.

Next, the *y*-coordinates of two horizontal lines are to be decided as follows:

$$seg\_line\_1 = y\_cent - seg\_height \text{ and } seg\_line\_2 = y\_cent + seg\_height$$

The portion/part of the image contained within these two horizontal lines is the desired segment to be extracted. The size of the extracted portion is half the height of the image. In terms of total number of pixels, the savings may not be half of the original image (generally).

But a major portion of the top one-fourth of the image containing upper body portion (e.g., head, shoulder etc.) and bottom one-fourth containing lower body portion (e.g., legs) are discarded from further consideration of feature extraction and classification.

### 4.11.3  Feature Extraction

There is a large number of shape descriptors, namely Fourier descriptor, width vector, projection profile, shape signature, and so on. A study is made using CASIA-A dataset (for three views) for finding out which among the four descriptors contain most discriminating features. Principal PCA is done on the set of feature vectors generated by the above-mentioned shape descriptors. The number of principal components that accumulate 95% variance is found out.

It is evident that the Fourier descriptor is the weakest among all the shape descriptors. The width vector and projection profile go hand in hand. The shape signature is proven to be the most effective shape descriptor. Being the strongest shape descriptor, the shape signature is used in this work for extracting features.

Signature method is a well-known image boundary descriptor [47]. The signature is a 1D representation of the boundary of an object. One of the simplest ways to represent it is to plot the distance un-warping the points lying on the boundary of an object from an interior point (e.g., centroid) as a function of angle.

In the present method, only the left and right boundaries of the segmented silhouette are considered for feature extraction. Therefore, the distances of all the left and right boundary pixels lying only on the segmented portion of a size-normalized silhouette are computed from the centroid of the object. The distances are considered from a fixed starting point covering all the left and right boundary points tracing the boundary in a counterclockwise direction. These distance vectors retain the shape features of the human silhouettes.

Signature generated by this approach is invariant to translation, but it depends on rotation. Normalization with respect to rotation can be achieved by finding a way to select the same starting point to generate the signature all the time, regardless of the orientation of the shape. One way to do so is to select the same starting point whenever calculating the signature. There are several possibilities of selecting the same starting point. The starting point could be the point farthest from the origin on the boundary of an object. Alternatively, a major axis can be computed with respect to the object boundary. The point of intersection of the major axis with the boundary of the object can be taken as the starting point for calculating the shape signature. Scale invariance is obtained by normalizing the distance from the centroid.

### 4.11.4 Classification

The system consists of two phases: training and testing. It is required to identify the landmark frames representing the double-support and mid-stance poses in a gait cycle. Training and testing are based only on the selected landmark frames.

### 4.11.5 Training

In the training phase the landmark frames are identified. Subimages are segmented out from the selected frames. Feature vectors are generated using shape signature as described in the previous subsection. The feature vectors are considered for PCA for dimension reduction of the vectors to a lower subspace. The PCA analyzed vectors are fed for multiple discriminant analysis (MDA) for further reduction of feature space to a single dimension for generating the template vectors. The set of all template vectors are kept to be used during testing phase.

*Feature dimension reduction*: Some dimensions may be dependent on other dimensions; therefore, those dimensions can be reduced. This can be done by PCA. In PCA the high dimensional feature vectors are projected to a lower dimensional eigenspace.

Given $s$ classes for training, and each class represents a sequence of distance signals of one subject's gait. Multiple sequences of each person can be freely added for training. Let $D_{ij}$ be the $j$th distance signal in class $i$ and $N_i$ is the number of such distance signals in the $i$th class. The total number of training samples $N_t = N_1 + N_2 + \cdots + N_s$ and the whole training set can be represented by $[D_{11}, D_{12}, \ldots, D_{1N_1}, D_{21}, \ldots, D_{2N_2} \ldots, D_{s1} \ldots, D_{sN_s}]$. The mean and the global covariance matrix of this set of training distance signals can be calculated as follows

$$m_d = \frac{1}{N_t} \sum_{i=1}^{s} \sum_{j=1}^{N_i} D_{ij} \quad \text{and} \quad -\sum = \frac{1}{N_t} \sum_{i=1}^{s} \sum_{j=1}^{N_i} \left(D_{ij} - m_d\right)\left(D_{ij} - m_d\right)^T$$

If the rank of the matrix is $N$, then $N$ nonzero eigenvalues $\lambda_1, \lambda_2 \dots \lambda_N$ and associated eigenvectors $e_1, e_2, \dots, e_N$ can be computed. The first few eigenvectors correspond to large change in training patterns. So, only $k$ out of $N$ eigenvectors that accumulates 95% variance are chosen and first $k$ eigenvectors are considered for dimension reduction.

Taking only the $k < N$ largest eigenvalues and their associated eigenvectors, the transform matrix $E = [e_1, e_2, \dots, e_k]$ is constructed to project an original distance signal $D_{ij}$ into a point $P_{ij}$ in the $k$-dimensional eigenspace. $P_{ij} = [e_1, e_2, \dots, e_k]^T D_{ij}$.

It is well-known that $k$ is usually much smaller than the original data dimension $N$. That is to say, eigenspace analysis can drastically reduce the dimensionality of input feature vectors. The final matrix consists of the final feature vectors of all the $s$ classes, each of the classes consisting of $N_i$ feature vectors, $i = 1$ to $s$.

*Generation of template vectors*: Projected vectors of each class are further considered for MDA to reduce the set of vectors of each class to a single dimension template vector.

The mean vector of all the feature vectors of each class is computed as follows:

$$m_i = \frac{1}{N_i} \sum_{j=1}^{N_i} P_{ij}$$

The global mean of all the feature vectors contained in the training set is computed as

$$m = \frac{1}{N_t} \sum_{i=1}^{s} \sum_{j=1}^{N_i} P_{ij}$$

The scatter matrix for each class is computed as

$$S_i = \sum_{j=1}^{N_i} \left(T_{ij} - m_i\right)^T \left(T_{ij} - m_i\right)$$

After this the within-class scatter matrix is calculated as follows:

$$S_w = \sum_{i=1}^{s} S_i$$

Next, the between-class scatter matrix is computed as follows:

$$S_B = \sum_{i=1}^{C} N_i \left( m_i - m \right)^T \left( m_i - m \right)$$

Now the eigenvectors of $(S_w)^{-1}S_B$ is calculated. No more than $C$-1 eigenvectors can be obtained. The eigenvectors are $[v_1, v_2,..., v_{C-1}]$.

The final feature vectors (templates) are obtained from previous feature vectors $P$ projected by PCA as follows: $F = P * [v_1, v_2,..., v_{C-1}]$.

### 4.11.6 Testing

In the testing phase, similar series of tasks are performed for feature dimension reduction. A minimum distance classifier based on Euclidean distance is used for testing. The signature of a test sample is compared with all the template signatures (each one is a candidate from a particular training class). The smallest distance represents the closest match. The test sample is considered to be in the same class as that of the template of the closest match.

Training and testing samples, for most of the cases, are taken from two different recording sessions of the same subject. In some cases, the training and testing samples are taken from the same recording session however from two different walking cycles.

The effectiveness of the partial silhouette-based method is tested on two publicly available gait datasets, namely, CASIA-A [12] and CMU MoBo [46].

### 4.12 Experimental Verification

The effectiveness of the partial silhouette-based method is verified for two situations with varying viewing angles of the camera (e.g., lateral view, frontal view, oblique view, etc. and CASIA-A dataset) and different walking speeds of the subjects (fast vs. slow walk, etc., CMU MoBo dataset).

*4.12.1 Results of Full versus Partial Silhouettes*

This work considers both silhouettes of the whole object (represented as full silhouette in the tables) as well as the partial silhouettes of human body having the portion of the body containing hand dynamics.

The comparative results of full versus partial silhouette for three different viewing angles (lateral/frontal/oblique) on CASIA Dataset-A is presented in Table 4.1. Results collected in Table 4.1 show the effectiveness of the partial silhouettes over full silhouettes of the subjects.

The lateral view is most suited for silhouette-based gait analysis. Eighty-five percent accuracy is obtained with partial silhouette opposed to 80% using full silhouette for CASIA-A in lateral view. Recognition accuracy for full silhouette falls because of the dissimilarities in the shapes of the silhouettes in the regions of legs and upper portion of the body. The lower portions of silhouettes sometimes contain shadow regions due to incorrect object segmentation of the background subtraction step required to generate the silhouettes from video.

Frontal view contains relatively less information for silhouette-based gait analysis. However, the partial silhouette method also wins over full silhouette in frontal views too. In oblique view full silhouettes performs better than partial silhouette. In this dataset the size of the silhouettes are relatively smaller. In the oblique view, the subjects walk from a distant location toward the camera in an oblique path. The partial silhouettes for training and testing when the subjects are far distant apart from the camera do not contain sufficiently discriminating information for human identification. This is why the full silhouette wins over partial silhouette-based method in oblique view.

**Table 4.1**  Results on CASIA Dataset-A

| TRAIN/TEST CONFIGURATION | RECOGNITION ACCURACY (%) | |
| --- | --- | --- |
| | FULL SILHOUETTE | PARTIAL SILHOUETTE |
| Lateral/lateral | 80 | 85 |
| Frontal/frontal | 70 | 75 |
| Oblique/oblique | 75 | 60 |

The experimentation results with CMU MoBo dataset [46] is presented in Table 4.2. Different training/testing combinations are considered for fast and slow walks for three different views (view-1 lateral, view-3 frontal, and view-5 oblique).

In lateral view, the partial silhouette-based method results in 88% accuracy compared to 76% of the full silhouette when testing samples in fast walking type is compared with the same type of training samples. In a slow/slow training/testing combination, 88% and 78% accuracy is achieved for partial and full silhouettes, respectively, in the same view. Other results for different walking types (fast/slow) and different views (frontal/oblique) are also presented in Table 4.2.

In frontal view, both partial and full silhouette method achieve 96% accuracy in fast/fast configuration. In slow/slow configuration the partial silhouette-based method performs better producing 96% accuracy than 92% produced by full silhouette. In oblique view for both fast/fast and slow/slow configurations the partial silhouette-based method produces 88% recognition accuracy.

Recognition accuracies for all the views fall significantly in fast/slow and slow/fast training/testing configurations. The reason for this is that there is a significant change in the human silhouette as a person walks in a fast motion compared to when the same person walks slowly. The body dynamics especially the swinging of hands and the legs change significantly. This is reflected in the sharp decrease

**Table 4.2**  Results on CMU MoBo Dataset

| CAMERA VIEWS | TRAIN/TEST CONFIGURATION | RECOGNITION ACCURACY (%) | |
|---|---|---|---|
| | | FULL SILHOUETTE | PARTIAL SILHOUETTE |
| View-1 lateral | Fast/fast | 76 | 88 |
| | Slow/slow | 78 | 88 |
| | Fast/slow | 16 | 28 |
| | Slow/fast | 16 | 32 |
| View-3 frontal | Fast/fast | 96 | 96 |
| | Slow/slow | 92 | 96 |
| | Fast/slow | 24 | 24 |
| | Slow/fast | 24 | 28 |
| View-5 oblique | Fast/fast | 88 | 88 |
| | Slow/slow | 84 | 88 |
| | Fast/slow | 12 | 16 |
| | Slow/fast | 16 | 36 |

of recognition accuracy when training samples are from fast walking types and the testing samples are from slow walking types or vice versa.

## 4.13 Comparison with Other Methods

This section presents a comparative study on a number of existing gait recognition techniques versus the partial silhouette-based method. Table 4.3 presents the results obtained when the partial silhouette-based method is compared with three different methods proposed by Kale et al. [10], Collins et al. [30], and Murat and Murat [44] on CMU MoBo dataset for slow/slow and fast/slow training/testing configurations in lateral view.

Results show that in lateral view for slow/slow configuration the method proposed by Murat and Murat [44] produces 95% accuracy followed by partial silhouette-based method achieving 88% at rank-1. However, the method by Murat and Murat [44] is expensive in terms of computation time and memory requirements compared to the partial silhouette-based method. Performance of the method proposed by Collins et al. [30] stands third giving 86%. The method proposed by Kale et al. [10] achieves 72% recognition accuracy at rank-1. As seen from the results presented in Table 4.4 that the performance of the partial silhouette-based method is second best compared to all.

The partial silhouette-based method is also compared with six other methods for gait recognition on CASIA-A dataset. The results are summarized in Table 4.4.

At Rank-1 the method proposed by Murat and Murat [44] performs the best followed by Lee and Grimson [31]. At Rank-1 the partial silhouette-based method gives the third best results (85%). However, in terms of computational complexity, partial silhouette-based

**Table 4.3**  Comparison of Different Methods on CMU MoBo Dataset (View 1: Lateral View)

| METHODS | TRAIN/TEST CONFIGURATION | RECOGNITION ACCURACY (%) | | | |
|---|---|---|---|---|---|
| | | RANK 1 | RANK 2 | RANK 3 | RANK 5 |
| Kale et al. [10] | Slow/slow | 72 | 80 | 85 | 97 |
| Collins et al. [30] | Slow/slow | 86 | 100 | 100 | 100 |
| Murat and Murat [44] | Slow/slow | 95 | – | – | – |
| Partial Silhouette [48] | Slow/slow | 88 | 96 | 100 | 100 |

**Table 4.4** Comparison of Different Methods on CASIA-A Dataset (View 1: Lateral View)

| METHODS | RECOGNITION ACCURACY (%) | | | | |
|---|---|---|---|---|---|
| | RANK 1 | RANK 2 | RANK 3 | RANK 5 | RANK 10 |
| BenAbdelkader et al. [22] | 73 | – | – | 89 | 96 |
| Collins et al. [30] | 71 | – | – | 79 | 88 |
| Lee and Grimson [31] | 88 | – | – | 99 | 100 |
| Phillips [25] | 79 | – | – | 91 | 99 |
| Wang et al. [12] | 75 | – | – | 98 | 100 |
| Murat and Murat [44] (Fusion) | 89 | – | – | – | – |
| Partial Silhouette [48] | 85 | 88 | 90 | 100 | 100 |

method is less expensive compared to these two methods. The partial silhouette-based method achieves 100% accuracy at Rank-5 giving better results than rest of the five methods (BenAbdelkader et al. [22], Collins et al. [30], Lee and Grimson [31], Sarkar [25]. Wang et al. [12]). This shows that partial silhouette-based method converges early compared to other methods. The entries marked by hyphens in the table represent that the corresponding values are not reported in the literature.

### 4.14 Effectiveness of Partial Silhouette Method in the Presence of Noise

All silhouette-based approaches depend on generation of a good quality silhouette of the moving objects. Most of the methods [12,44] including the partial silhouette-based method, depend on the description of only the boundary of the silhouette. Thus, it is required to extract the silhouette as efficiently as possible.

Video data is captured using a camera, and frames are extracted from them. Generally, the background subtraction techniques are used for extracting moving objects and generating the binary silhouettes [12]. It has been observed that when working with real-life video, the extracted binary silhouettes contain noises at the bottom portions of the frames due to shadows. This results in imperfect or incomplete silhouettes generated by the background subtraction method. There are various other types of noises present in real-life videos due to sudden illumination changes in the scene, camera sensor errors, occlusions, and so on.

The partial silhouette-based method is effective when the silhouettes of the objects contain imperfections in terms of noises. The recognition accuracy of the partial silhouette-based method does not get much affected if the portion of the silhouette containing the regions of hands can be extracted properly.

Figure 4.11 shows a set of six noisy silhouette images. The image in Figure 4.11b suffers from shadow-related noise. Other silhouettes are incomplete and imperfect due to presence of various kinds of noises described earlier.

A gait recognition technique based on full object silhouette will fail to detect these noisy samples if the training samples are not trained to handle such imperfections. However, the partial silhouette-based method will correctly classify some of the samples (Figure 4.11b, d, and e).

The segmented silhouettes on which the partial silhouette-based method will work are shown in Figure 4.12 for silhouettes (b), (d), and (e) of Figure 4.11. It is to be noted that the portions of the segmented regions do not contain any major noise. Thus, partial signatures generated from these segmented silhouettes will correctly identify the subjects.

Figure 4.13 shows the signatures of the template samples corresponding to the silhouettes shown in Figure 4.11a, c, and f in the first

**Figure 4.11** (a–f) A set of noisy silhouettes.

**Figure 4.12** Segmented silhouettes of noisy images (from left to right) of Figure 4.11 (b), (d), and (e), respectively.

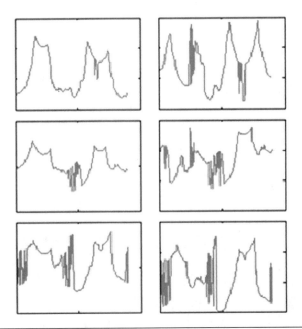

**Figure 4.13** Signatures of silhouettes of samples (a), (c), and (f) shown in Figure 4.11.

column. In the second column the signatures of the partial silhouettes of the testing samples are given.

As seen from this figure, there are significant changes in the signatures of the training and testing samples leading to misclassifications.

Similarly, Figure 4.14 presents the training and testing signatures for samples b, d, and e of Figure 4.11. It is clear from Figure 4.14 that there are significant similarities between the signatures of the training and that of the testing samples. This shows the partial silhouette-based method is more noise-immune in such situations.

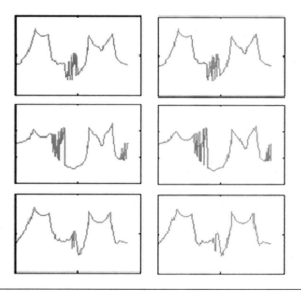

**Figure 4.14** Signatures of silhouettes of samples (b), (d), and (e) shown in Figure 4.11.

## 4.15 Time Complexity of the Partial Silhouette-Based Method

The new method is a low-cost solution for human gait recognition. A portion of the human silhouette is extracted from the whole binary silhouette image. Human recognition is carried out by extracting features from this extracted portion of the silhouette. This results in handling only a part of the binary silhouette rather than considering the whole silhouette of a moving subject for feature extraction. There are other approaches [12,32,44] that rely on extracting features from the boundary of whole object silhouette or most part of it. Existing methods (Fourier descriptor [32], GEI [26], boundary shape signature [12], projection-based method [44]) depend on the set of all frames in a gait cycle. Training of the classifier requires handling huge data. This is expensive in terms of computations.

The time complexity is related to the size of the feature vector a method has to handle during training and testing phases. We consider that the silhouettes under observation are size-normalized to $N$ scan lines for height. A conventional silhouette-based method will use a feature vector of dimension at least $2N$ per frame. If the feature is based on the whole object boundary as in Fourier descriptor or shape signature, the dimension of a feature vector will be even more. If the average number of frames per cycle be $M$, then the system has to handle feature vectors of dimension $2*N*M$ per gait cycle for training. This is typically the case of methods that use all the frames of a cycle during training.

In the partial silhouette-based method, training and testing are done based on landmark frames only (only 4 landmark frames per gait cycle). Moreover, silhouette is segmented to 1/2 of the normalized size $N$ resulting in $4*N$ sized data per gait cycle. This shows that in terms of computational complexity, the partial silhouette-based method is light weight compared to other methods.

Space requirement is another issue. Template vectors of dimensions $2N$ have to be stored for other vision-based approaches. However, the partial silhouette-based method stores information about only half of the silhouette. It considers both left and right contours of the silhouette totaling a vector of dimension of $N$ resulting in 50% savings in storage space compared to other methods.

On an average, the method proposed by Wang et al. [12] is arguably one of the most widely used methods for silhouette-based gait recognition. This method, in terms of accuracy detects 75% of the cases for the given dataset. In contrast, the partial silhouette-based method performs better as documented in Table 4.4. Moreover, lower time and space complexities of the partial silhouette-based method make it suitable for a plethora of applications. As mentioned above, the partial silhouette-based method can be used on its own as a human verification methodology, particularly in small groups of people. However, in large groups of users, where the feature repeatability is essential, the partial silhouette-based method can be used in a system employing multimodal biometrics where gait is one of the modalities. The overall accuracy of the system can then be increased by other biometric modalities like face recognition, fingerprint, or other possible physiological biometrics features used for human recognition.

The partial silhouette-based methodology proposed in [48] has been documented in [49,50]. This method is shown to be performing well in the presence of certain kind of noise. The comparative study of this method with some of the existing gait recognition techniques using two public gait datasets reveals the effectiveness of this approach. The low-cost nature of the method makes it suitable to be incorporated in multimodal security systems as well.

## 4.16 Conclusions

Gait analysis may be employed for verifying the identity of a person without resort to documents that may be stolen, lost, or altered. Gait recognition is attributed with nonobtrusiveness, noninvasiveness, and recognition at a distance. There are several challenges that exist for real-life gait recognition. Low repeatability is a common problem in all behavioral biometric approaches including vision-based gait analysis. There are other internal and external factors that result in poor recognition accuracy for gait analysis. Recent trends show that researchers are more interested in multimodal biometric systems rather than focusing on solving a particular type of challenge involved in gait analysis. Combination of different biometric technologies, for example, gait along with face recognition, gait along with ear biometrics, and so on have been widely used by the researchers in search of a robust system.

There are a number of applications of gait analysis. Vision-based gait analysis is primarily used in security and surveillance applications. Sensor-based gait analysis has found its applications in medical diagnosis.

Benchmark databases are important for experimental analysis of an algorithm. This chapter also summarizes some databases available for gait analysis. In this chapter, a partial silhouette-based method for gait analysis has been discussed. This method considers only a portion/part of the whole human silhouette containing the region of swinging hands. This region of human body contains sufficient discriminating information for gait analysis. This method is low cost in terms of both computation time and memory requirements as data related to only the partial-silhouette has to be dealt with during feature extraction and subsequent classification phases. This method also performs well in the presence of certain type of noise. The effectiveness of this method is verified on two benchmark databases.

# References

1. N. V. Boulgouris, K. N. Plataniotis, E. Micheli-Tzanakou (eds.), *Biometrics: Theory, Methods, and Applications*, John Wiley & Sons, Hoboken, NJ, 2009.
2. R. V. Yampolskiy, V. Govindaraju, Behavioural biometrics: A survey and classification. *International Journal of Biometrics*, 1(1), 81–113, 2008.

3. W. H. Abdulla, Y. Zhang, Voice biometric feature using Gammatone Filterbank and ICA, in special issue Speech as a human biometric: I know who you are from your voice! *International Journal of Biometrics*, 2(4), 330–349, 2010.

4. W. H. Abdulla, P. Yu, P. Calverly, Lips tracking biometric for speaker recognition. *International Journal of Biometrics*, 1(3), 288–306, 2009.

5. N. Nishiuchi, S. Komatsu, K. Yamanaka, Biometric verification using the motion of fingers: A combination of physical and behavioural biometrics. *International Journal of Biometrics*, 2(3), 222–235, 2010.

6. R. V. Yampolskiy, M. L. Gavrilova, Artimetrics: Biometrics for artificial entities. *IEEE Robotics & Automation Magazine*, 19(4), 48–58, 2012.

7. A. W. Mitas, M. D. Bugdol, Strengthening a cryptographic system with behavioural biometric. In *3rd International Conference Information Technologies in Biomedicine*, Springer, Gliwice, Poland, pp. 266–276, 2012.

8. M. Godzwon, K. Saeed, Biometrics image denoising algorithm based on Contourlet transform. In *International Conference on Computer Vision and Graphics*, Springer, Warsaw, Poland, pp. 735–742, 2012.

9. J. Wang, M. She, S. Nahavandi, A. Kouzani, A review of vision-based gait recognition methods for human identification. In *Proceedings of the International Conference on Digital Image Computing: Techniques and Applications*, IEEE, Sydney, Australia, 2010.

10. A. Kale, A. Sundaresan, A. N. Rajagopalan, N. P. Cuntoor, A. K. Roy-Chowdhury, V. Kruger, R. Chellappa, Identification of humans using gait. *IEEE Transactions on Image Processing*, 13(9), 1163–1173, 2004.

11. Z. Liu, S. Sarkar, Outdoor recognition at a distance by fusing gait and face. *Image and Vision Computing*, 25(6), 817–832, 2007.

12. L. Wang, T. Tieniu, N. Huazhong, H. Weiming, Silhouette analysis-based gait recognition for human identification. In *IEEE Transactions on Pattern Analysis and Machine Intelligence*, 25, 1505–1518, 2003.

13. C. Bauckhage, J. K. Tsotsos, F. E. Bunn, Automatic detection of abnormal gait. *Image and Vision Computing*, 27(1–2), 108–115, 2009.

14. L. Xuelong, S. Maybank, T. Dacheng, Gender recognition based on local body motions. In *IEEE International Conference on Systems, Man and Cybernetics*, IEEE, Montreal, Canada, pp. 3881–3886, 2007.

15. L. Xuelong, S. J. Maybank, Y. Shuicheng, T. Dacheng, X. Dong, Gait components and their application to gender recognition. *IEEE Transactions on Systems, Man, and Cybernetics, Part C: Applications and Reviews*, 38, 145–155, 2008.

16. Y. Iwashita, K. Uchino, R. Kurazume, Gait-based person identification robust to changes in appearance. *Sensors*, 13(6), 7884–7901, 2013.

17. Y. Guan, C. T Li, A robust speed-invariant gait recognition system for Walker and Runner identification. In *6th IAPR International Conference on Biometrics*, IEEE, Madrid, Spain, pp. 1–8, 2013.

18. Y. Guan, C.-T. Li, Y. Hu, Robust clothing-invariant gait recognition. In *International Conference on Intelligent Information Hiding and Multimedia Signal Processing*, IEEE, Piraeus, Greece, pp. 321–324, 2012.

19. M. Hofmann, J. Geiger, S. Bachmann, B. Schuller, G. Rigoll, The TUM gait from audio, image and depth (GAID) database: Multimodal recognition of subjects and traits. *Journal of Visual Communication and Image Representation*, 25(1), 195–206, 2014.

20. E. Hossain, G. Chetty, R. Goecke, Multi-view multi-modal gait based human identity recognition from surveillance videos. In *Multimodal Pattern Recognition of Social Signals in Human-Computer-Interaction*, Springer, Tsukuba, Japan, pp. 88–99, 2012.

21. R. Zhang, C. Vogler, D. Metaxas, Human gait recognition at sagittal plane. *Image and Vision Computing*, 25(3), 321–330, 2007.

22. C. BenAbdelkader, R. Cutler, L. Davis, Stride and cadence as a biometric in automatic person identification and verification. In *5th IEEE International Conference on Automatic Face and Gesture Recognition*, IEEE, Washington, DC, pp. 372–377, 2002.

23. A. F. Bobick, A. Y. Johnson, Gait recognition using static, activity-specific parameters. *IEEE Computer Society Conference on Computer Vision and Pattern Recognition*, 1, 423–430, 2001.

24. N. V. Boulgouris, D. Hatzinakos, K. N. Plataniotis, Gait recognition: A challenging signal processing technology for biometric identification. *IEEE Signal Processing Magazine*, 22(6), 78–90, 2005.

25. S. Sarkar, P. J. Phillips, Z. Liu, I. R. Vega, P. Grother, K. W. Bowyer, The humanID gait challenge problem: Data sets, performance, and analysis. *IEEE Transaction on Pattern Analysis and Machine Intelligence*, 27(2), 162–177, 2005.

26. J. Han, B. Bhanu, Individual recognition using gait energy image. *IEEE Transactions on Pattern Analysis and Machine Intelligence*, 28(2), 316–322, 2006.

27. L. Jianyi, Z. Nanning, Gait history image: A novel temporal template for gait recognition. In *International Conference on Multimedia and Expo*, IEEE, Beijing, China, pp. 663–666, 2007.

28. C. Chen, J. Liang, H. Zhao, H. Hu, J. Tian, Frame difference energy image for gait recognition with incomplete silhouettes. *Pattern Recognition Letters*, 30(11), 977–984, 2009.

29. P. Phillips, S. Sarkar, I. Robledo, P. Grother, K. Bowyer, Baseline results for the challenge problem of human ID using gait analysis. In *5th International Conference on Automatic Face and Gesture Recognition*, IEEE, Washington, DC, 2002.

30. R. Collins, R. Gross, J. Shi, Silhouette-based human identification from body shape and gait. In *International Conference on Automatic Face and Gesture Recognition*, IEEE, Washington, DC, 2002.

31. L. Lee, W. Grimson, Gait analysis for recognition and classification. In *IEEE International Conference on Automatic Face and Gesture Recognition*, IEEE, Washington, DC, pp. 155–162, 2002.

32. T. Guang-Jian, H. Fu-Yuan, Z. Rong-chun, Gait recognition based on Fourier descriptors. In *International Symposium on Intelligent Multimedia, Video and Speech Processing*, IEEE, Dallas, TX, pp. 29–32, 2004.

33. G. V. Veres, L. Gordon, J. N. Carter, M. S. Nixon, What image information is important in silhouette-based gait recognition? In *IEEE Computer Society Conference on Computer Vision and Pattern Recognition*, IEEE, Washington, DC, 2004.

34. N. Liu, Y.-P. Tan, View invariant gait recognition. In *IEEE International Conference on Acoustics Speech and Signal Processing*, IEEE, Dallas, TX, pp. 1410–1413, 2010.

35. A. P. Yazdanpanah, K. Faez, R. Amirfattahi, Multimodal biometric system using face, ear and gait biometrics. In *10th International Conference on Information Sciences Signal Processing and Their Applications*, IEEE, Kuala Lumpur, Malaysia, pp. 251–254, 2010.

36. D. Kim, J. Paik, Gait recognition using active shape model and motion prediction. *IET Computer Vision*, 4(1), 25–36, 2010.

37. S. Sharma, R. Tiwari, A. Shukla, V. Singh, Fusion of gait and facial feature using PCA. *Signal Processing, Image Processing and Pattern Recognition*, 260, 401–409, 2011.

38. E. Hossain, G. Chetty, Multimodal identity verification based on learning face and gait cues. *18th International Conference on ICONIP*, 7064, 1–8, 2011.

39. M. Hu, Y. Wang, Z. Zhang, Z. Zhang, Multi-view multi-stance gait identification. In *18th IEEE International Conference on Image Processing*, IEEE, Brussels, Belgium, pp. 541–544, 2011.

40. I. Venkat, P. D. Wilde, Robust gait recognition by learning and exploiting sub-gait characteristics. *International Journal of Computer Vision*, 91(1), 7–23, 2011.

41. R. Martin-Felez, T. Xiang, Gait recognition by ranking. In *12th European Conference on Computer Vision, Part I*, Springer, Florence, Italy, pp. 328–341, 2012.

42. N. V. Boulgouris, Z. X. Chi, Human gait recognition based on matching of body components. *Pattern Recognition*, 40(6), 1763–1770, 2007.

43. G. Baofeng, M. S. Nixon, Gait feature subset selection by mutual information. In *IEEE Transactions on Systems, Man and Cybernetics, Part A: Systems and Humans*, 39(1), 36–46, 2009.

44. E. Murat, A. Murat, Improved gait recognition by multiple-projections normalization. *Springer Journal on Machine Vision and Applications*, 21(2), 143–161, 2010.

45. A. Hossain, Y. Makihara, J. Wang, Y. Yagi, Clothing invariant gait identification using part-based clothing categorization and adaptive weight control. *Pattern Recognition*, 43(6), 2281–2291, 2010.

46. R. Gross, J. Shi, The CMU Motion of Body (MoBo) database. Tech. Report CMU-RI-TR-01-18, Robotics Institute, Carnegie Mellon University, 2001.

47. R. Gonzalez, R. Woods, *Digital Image Processing*, International 3rd Revised Edition, Chapter 10, Pearson Prentice Hall, Upper Saddle River, NJ, 2008.

48. S. H. Shaikh, Image binarization and analysis towards moving object detection and human gait recognition, PhD Thesis. Department of Computer Science and Engineering, University of Calcutta, India, 2014.

49. S. H. Shaikh, K. Saeed, N. Chaki, Gait recognition using partial silhouette-based approach. In *IEEE Proceedings of the International Conference on Signal Processing & Integrated Networks*, IEEE, Noida, India, 2014.

50. S. H. Shaikh, K. Saeed, N. Chaki, Partial silhouette-based gait recognition. *International Journal of Biometrics*, 2016 (in press).

# 5

# VOICE RECOGNITION

This chapter explains fundamental concepts of voice recognition necessary for the reader to understand the mentioned techniques, methods, and algorithms. It also contains the main steps related to the applied solutions in voice, speech, and speaker recognition systems.

## 5.1 Voice Recognition

In modern cultures, people use technical solutions, combined with smart instruments, mobiles, and web devices almost every minute, where the usability and accessibility are fundamental parts of today's applications and needs. Thus, the demand for voice recognition has increased; the fast-paced world also increases the need for integrated solutions. People should know how to handle their daily tasks by voice commands, as the related consequences bind identities depending on who is saying and what is being said.

From scientific point of view, voice recognition is a field of computer science that depends on computer software and algorithms to identify different features and properties related to speech and speakers. Usually, voice recognition can be divided mainly into speech and speaker recognition—identification or verification.

The starting point for these tasks is the speech signal. Speech is a complicated signal produced as a result of several transformations occurring at semantic, linguistic, articulatory, and acoustic levels. Differences in these transformations appear as differences in the acoustic properties of the speech signal. Speaker-related differences are a result of a combination of anatomical differences inherent in the vocal tract and the acquired speaking habits of different individuals. In speaker recognition, all these differences can be used to discriminate between speakers.

Automatic speaker verification (ASV) and automatic speaker identification (ASI) are probably the most natural and economical methods for solving the problems of unauthorized use of computer and communications systems and multilevel access control. With the ubiquitous telephone network and microphones bundled with computers, the cost of a speaker recognition system might only be for software.

Biometric systems automatically recognize a person by using distinguishing traits. Speaker recognition is a performance biometric, that is, user performs a task to be recognized. User's voice, like other biometrics, cannot be forgotten or misplaced, unlike knowledge-based (e.g., password) or possession-based (e.g., key) access control methods [1].

Two major possibilities are included in the speaker recognition task: first, the ASI, which involves detecting the person who speaks from a certain group of speakers. In its ideal situation, this task can be made independently from the uttered speech; easier assignment can be done with certain words or phrases to be spoken.

The second possibility is the ASV, which involves the use of algorithms to verify the identity of a person from his voice. This kind of verification is combined with authentication systems and has a high number of applications and useful practical sides.

On the other hand, speech recognition refers to understanding what the voice is, and what kind of information it contains. For example, what are the phrases and sentences? In which languages is it? And what is the emotional case of the speaker? Some of these systems rely on certain speakers, where the training process focuses on a certain speaker to increase accuracy, in such a case, this kind of software usually can handle an open set of vocabulary during the recognition process. Meanwhile, other systems are handling the case of speech recognition with an independent speaker (open set or close set of vocabulary).

*5.1.1 Advantages of Voice Recognition over Other Biometric Traits*

Biometric systems can bind human identity to electronic series of features. This task can be done using different kinds of human traits like fingerprints, iris, or voice.

**Table 5.1** Comparison Biometric Systems Based on Biometric Characteristics

| BIOMETRIC SYSTEM | SIMPLICITY | COST | EMOTIONAL STABILITY | AGE STABILITY | ACCURACY |
|---|---|---|---|---|---|
| Fingerprints | + | ± | + | + | + |
| Hand | ± | ± | ± | − | ± |
| Iris | − | − | + | + | + |
| Face | − | + | ± | − | + |
| Signature | ± | ± | + | + | + |
| Voice | + | + | ± | − | ± |

*Note:* Ratings are indicated as negative (−), fair (±), or positive (+).

In order to clarify the importance of voice as an identifier for human beings, we will present a simple table with the most human traits used in biometric systems (Table 5.1).

### 5.1.2 Main Steps in Voice Recognition Systems

There are certain common steps usually used in voice recognition systems. These steps are presented in Figure 5.1.

As explained in Figure 5.1, the standard voice recognition system consists of four main steps: signal acquisition, preprocessing, feature extraction, and classification.

## 5.2 Signal Acquisition and Preprocessing

Signal acquisition takes the voice from the external source; usually it comes from microphone, where the voice can be recorded in the computer. In other cases authors used already recorded samples as PC files, making the software read these files directly from the local storage.

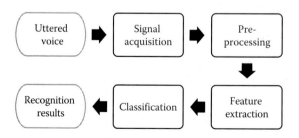

**Figure 5.1** Procedure diagram generally used in voice recognition systems.

In all these instances, the authors worked on simple hardware and software solutions, and did not rely on this step as in phonetics laboratories, meanwhile all computations are performed in MATLAB®.

### 5.2.1 Biological Background

For better understanding of voice recognition approaches, first we will present a biological background about the speech and hearing organs in the human body.

The human voice is unique because of the "physiological and behavioral aspects of speech production." The shape of the vocal tract in humans is what makes the voice unique. The vocal tract's location is depicted by the shaded area shown in Figure 5.2 [2].

Speech can be defined as waves of air pressure created by airflow pressed out of the lungs and going out through the mouth and nasal cavities. The air passes through the vocal folds (chords) via the path from the lungs through the vocal tract, vibrating them at different frequencies.

The vocal folds are thin muscles looking like lips, located at the larynx. At their front end they are permanently joined, and at the other end they can be open or closed. When the vocal folds are closed, they are expanded against each other, forming an air block. Air under pressure from the lungs can open that air block, pushing the vocal folds aside.

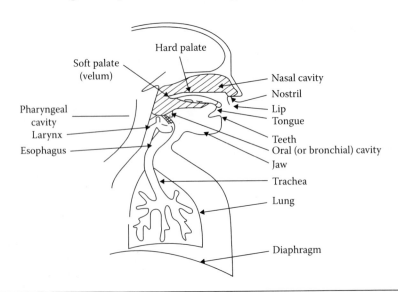

**Figure 5.2** Vital organs for speech and hearing.

The air passes through the crack thus formed and the pressure declines, allowing the vocal folds to close again. This process repeats, vibrating the vocal folds to give a voiced sound. The vocal folds in males are usually longer than in females, causing a lower pitch and a deeper voice.

When the vocal folds are open, they allow air to reach the mouth cavity easily. Unvoiced sound is formed by a constriction in the vocal tract, causing air turbulence and then random noise [3].

From the anatomical point of view the vocal tract consists of four cavities: larynx, annular cavity, oral cavity, and nasal cavity. These cavities contain members and each one of them has its role in the output of linguistic votes. These members work accurately and careful harmony exists between each other, controlled by nearly hundreds of muscles linked to the brain, and receive commands from the brain via the nerve networks that connect between these parts. As a summary, next are the different parts of the vocal tract:

- Larynx
- Pharyngeal cavity
- Nasal cavity
- Oral cavity
  - Lips and cheeks
  - Teeth
  - Tongue
  - Palate
  - Mandible [4]

Meanwhile, the physical structures of the ear are deceptively simple, especially in light of their exquisite function. The ear is an energy transducer, which means that it converts acoustic energy into electrochemical energy. The basic elements involved are outer ear, middle ear, inner ear, and auditory pathways.

The outer ear can be seen primarily as a collector of sound. The pinna, with its ridges, grooves, and dished-out regions, is an excellent funnel for sound directed toward the head from the front or side, although less effective for sound arising from behind the head.

The middle ear mechanism is designed to increase the pressure approaching the cochlea, thereby overcoming the resistance to flow of energy, termed impedance. The middle ear mechanism decreases the area over which the force is being exerted as the primary means of matching the impedance of the outer and inner ear. That is, the primary function of the middle ear is to match the impedance of two conductive systems, the outer ear and the cochlea.

In summary, the outer and middle ear serve as funneling and impedance-matching devices. Inner ear is responsible for performing spectral and temporal acoustic analyses of the incoming acoustical signal. By spectral analysis, we refer to the process of extracting or defining the various frequency components of a given signal. Frequency and intensity of vibration define the psychological correlates of "pitch" and "loudness." The cochlea is specifically designed to sort out the frequency components of an incoming signal, determine their amplitude, and even identify basic temporal aspects of that signal. These processes make up the first level of auditory processing of an acoustic signal. Subsequent processing occurs as the signal works its way rapidly along the auditory pathway, ultimately to the brain. To get a notion of how this happens, we need to consider the input to the cochlea.

Next come the auditory pathways, first organ in the auditory pathway is the cochlear nucleus where tonotopic representation is readily observable in tuning curves. There is evidence that significant signal processing occurs at this level of the brainstem. At least there are six different neural responses to auditory stimulation, in contrast to the single-unit response seen at the VIII nerve level. Primary-like responses are the firing patterns that most resemble VIII nerve responses. These responses appear to arise from bushy cells in the cochlear nucleus and will have identical rate functions, intensity responses, and spontaneous activity as the VIII nerve fibers they reflect [5].

### 5.2.2 Preprocessing Stage

The preprocessing stage deals with the input as a voice signal, applying its techniques on it, like different kinds of filters, bands, noise reduction, silence removing, and so on. After all these small modifications the voice signal will be ready for later usage; for recognition purpose, the signal will be ready to take its significant characteristics.

We handle the preprocessing stage in two different ways: (1) for uttered words recognition we deal with separated files. The input to the system is in a recorded speech waveform. Each file can contain one voice, or a chain of words and phrases depending on the experiment.

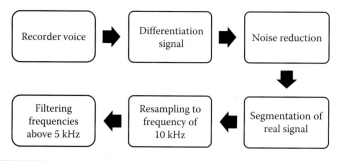

**Figure 5.3**   General steps of speech signal preparation for recognition process.

(2) In simple cases, there must be a silence region before and after the right signal. The speech preparation process for feature extraction is given in Figure 5.3.

Preprocessing stage depends on the goal of the tested algorithm, for example sometimes the authors decided to skip some of these steps to simplify the algorithm and testing its elasticity. Once these steps are finished, the speech signal is ready for the next step of feature extraction. The other case will be the preprocessing stage for text-independent speaker identification and verification, where the whole signal is in one file, and we divided it by taking only the parts with the most significant energy levels.

### 5.2.3  Feature Extraction

The feature extraction stage focuses on obtaining the most important voice characteristics, called features, these features should be as significant as most, and as few as possible, taking large number of features causes calculation difficulties for softwares, and makes it difficult to apply in real applications.

The voice signal was presented graphically using linear predictive coders, and Toeplitz matrix minimal eigenvalues algorithm, and other more simple ones; special try was given for standard audio algorithm for comparing purposes.

### 5.3  Toeplitz Matrix Minimal Eigenvalues Algorithm—A Survey

This algorithm attempt is continuous developing, meanwhile it achieved significant success with graphically illustrated signals for pattern recognition purposes. This algorithm has been tested in many experiments related to biometrics and has shown its successful performance in different fields [6].

In signal recognition, however, the work on Toeplitz matrix minimal eigenvalues theory has quickly been developed, although the success rate is still not as high as with that of object-image applications. This comes from two facts—the first lies in the complicated nature of voice and speech signals, and the second is the short age of the application of the theory of minimal eigenvalues to speech recognition [7]. Nevertheless, applying Toeplitz and radial neural networks (NNs) gave a 95.82% successful recognition for single Arabic words spoken by people selected from different countries (of different accents) at different ages and gender [8]. The main advantage of Toeplitz approach lies in its elasticity of fusing with other tools in a hybrid system [9].

Toeplitz matrix minimal eigenvalues algorithm reduces the amount of computations from operations on an $n \times n$ matrix that contains $n^2$ different elements to a matrix (of Toeplitz form) which contains only $n$ elements that are different from each other.

A brief discussion of the theory is given in [10–12]. Different modifications for TM such as modulus (absolute value), successive modulus differences, and polar representation are made [6,10]. More details about this approach and its applications can be found in [13].

### 5.3.1 Linear Predictive Coding and Burg's Model

Among many methods of speech signal processing, the authors have chosen the method based on spectrum analysis [14,15]. This method contributes to speech-image feature extract accomplishment by spectral analysis. Authors' experiments showed that the power spectrum estimation of Burg's model is one of the best methods for smoothing irregular spectral shape resulting from applying the fast Fourier transform (FFT) and the linear predictive coding approach [7,14,15].

Before entering this stage, the signal image needs some steps in speech preprocessing where Burg's model [16,17] (the frequency spectral estimation method, based on linear predictive coding principle [7,14]) is applied. Burg's model is built on the idea of prediction error minimization [14,15] and is explained in detail together with its software and computer implementation in [16].

Now we can use the obtained power spectrum acoustic images directly, or we can apply the algorithm based on minimal eigenvalues of Toeplitz matrices [6,7] to analyze these acoustic images. When using Burg's method of estimation, however, we need to specify the prediction order ($P$) and the FFT size, called the length of FFT (NFFT). The FFT length must give the smoothest shape of the spectrum (the more samples we have, the smoother shape we get), and it cannot be a case where too many samples are considered. This, as very well known, would definitely lower the efficiency of the algorithm. Prediction order ($P$) is also an important parameter. When it is too low, the envelope does not match with FFT shape, and when it is too high, it causes a speed falling of the algorithm. So it is very important, although very difficult, to choose the best prediction order. This had already been proved and shown in some previous work [7], where more explanations and details about Burg's method are explained. Toeplitz matrix minimal eigenvalues algorithm in its model for image description and feature extraction, however, is given in [6].

We have directed our previous researches on voice recognition [8] to the situation of recognizing the speaker himself. Before presenting the results we have achieved and the high efficiency of the implemented security system, we will first prove experimentally that the algorithm is valid for speaker identification as well as for his speech.

First, we present Burg's spectral graph of the Arabic word *sefer* (zero) uttered by three speakers (Figure 5.4).

We can see that each curve in Figure 5.4 has a specific shape differing from the other two, which makes the identification process theoretically possible [6,7]. One would recall all of the three graphs, from the other side, represent the same word, but spoken by three different people. One may ask, then how to distinguish between them, and the answer is simply in that although the shapes of these curves are almost similar and therefore they furnish a similar feature vector for the same word (sefer), which in turn shows characteristic behavior differing much from all the other digits' feature vectors, they (the curve shapes) belong to different speakers. This is because they differ from each other in only some features (coming from voice origin) that are never absolutely similar to each other or to other digits' characteristics. These almost similar features are the key for speaker identification.

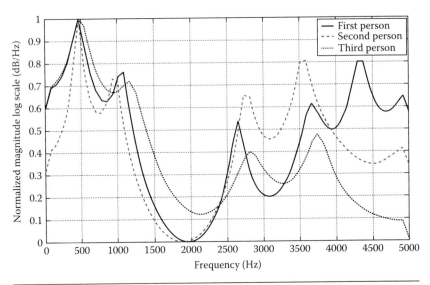

**Figure 5.4** Burg's curves for the Arabic word, *sefer* (one of the grammatically possible pronunciation of the digit, *zero*) spoken by three different people.

Therefore, the small differences between the curves in Figure 5.4, like the positions of the maxima, minima, and points of inflection, form the essence of the algorithm for distinguishing between the speakers' curves. The techniques followed in that are similar to those used in the graph presentation for speech (uttered words) recognition. The details are given in [7,18,19].

The acquired data from the curves of Figure 5.4, after computing the feature vector by Toeplitz matrix minimal eigenvalues algorithm [6], will enter the stage of classification (conventional or NN based). The following two sections will describe all the above shown issues demonstrating exactly how the system works.

### 5.3.2 Mel Frequency Cepstral Coefficients

The original Mel frequency cepstral coefficient (MFCC) algorithm introduced by Davis and Mermelstein (DM) [20] combined perceptually spaced filters with the discrete cosine transform (shown to be similar to principal eigenvectors of Dutch vowels) [21] in a mainstream speech processing publication. The algorithm can be summarized as follows: a time signal is windowed, FFT to the frequency domain, and scaled by a bank of triangular filters, equally spaced on a linear-log frequency axis, and the sum of magnitude coefficients scaled by each filter is log-compressed and transformed via the discrete cosine transform to cepstral coefficients. The filter bank is comprised of triangular filters, which are a coarse approximation to the shape of the critical bandpass response of the human auditory system. The base of each triangle is determined by the center frequencies of the adjacent filters; that is, filter bandwidth in MFCC is determined by the frequency range of the filter bank as well as the number of filters in the bank.

However, coupling bandwidth to other filter bank design parameters creates two problems. As automatic speech recognition (ASR) experimenters adapt the algorithm to their own desired frequency range (typically set by the sampling rate of the speech corpus under study), they unintentionally change filter bandwidth. MFCC came to the forefront of filter bank-based speech feature extractors because of the perceptually motivated filter spacing, yet the well-known relationship between frequency and critical bandwidth of the human auditory system is not incorporated. This shortcoming is rectified in human factor cepstral coefficients (HFCC). The second problem with coupling bandwidth to the number of filters and to frequency range is that filter bandwidth is not subject to optimization with respect to experimental recognition accuracy [22], meanwhile MFCC is widely used in voice applications [22,23].

## 5.4 Classification Using NNs

NNs stand as a very impressive tool for feature classification, the power and utility of the artificial NNs have been demonstrated in several applications including speech synthesis, robotic control, signal processing, computer vision, and many other problems related to the category of pattern recognition. Generally, different kinds of NNs have been tested by the authors showing promising results in achieving good performance over techniques of more traditional artificial intelligence character, especially when it comes to their use for the purpose of Arabic speech recognition [8,24]. Next in this section, we will present a brief theoretical consideration of the used NNs [25,26].

### 5.4.1 Probabilistic NNs

In the probabilistic NNs (PNNs) [25], there are at least three layers: input, radial, and output ones. The radial units are copied directly from the training data, one per case. Each of them models a Gaussian function centered at the training case. There is one output unit per class; each is connected to all the radial units belonging to its class, with zero connections from all other radial units. Hence, the output units simply add up the responses of the units belonging to their own class. The outputs are proportional to the estimates of the probability density functions of the various classes. The only control factor that needs to be selected for PNN training is the smoothing factor. This factor needs to be selected in such a way that it would only cause a reasonable portion of overlapping. An appropriate figure is easily chosen by experiment, by selecting a number which produces a low selection error, and fortunately PNNs are not too sensitive to the precise choice of smoothing factor.

The greatest advantage of PNNs is the fact that the output is probabilistic, which makes the interpretation of the output easier and the training speed higher [25]. Training a PNN actually consists mostly of copying training cases into the network, and so it is as close to the instantaneous value as can be expected. The greatest disadvantage is the network size: a PNN network actually contains the entire set of training cases, and is therefore space-consuming and slow to execute.

### 5.4.2 Radial Basis Function NNs

A radial basis function (RBF) network [26] has three layers: input, radial, and output layers. The hidden (radial) layer consists of radial units; each actually is modeling a Gaussian response surface. The units will always be sufficient to model any function. The RBF in each radial unit has a maximum of 1 when its input is 0. As the distance between the weight vector and the input decreases, the output increases. Thus, a radial basis neuron acts as a detector that produces 1 whenever the input is identical to its weight vector, additionally there is a bias neuron which allows the sensitivity of the radial basis transfer function of the hidden neurons to be adjusted. The standard RBF NN has an output layer containing dot product units with identity activation [26]. Radial basis networks may require more neurons than standard feed-forward back propagation networks, but often they can be designed in a fraction of the time it takes to train standard feed-forward networks. They work best when many training vectors are available. RBF networks have a number of advantages [26]. First, they can model any nonlinear function using a single hidden layer, which eliminates some design decisions about the number of layers. Second, the simple linear transformation in the output layer can be optimized fully using traditional linear modeling techniques, which are fast and do not pose problems, RBF networks can therefore be trained extremely quickly, training of RBFs takes place in distinct stages. The centers and deviations of the radial units must be set up before the linear output layer is optimized.

## 5.5 Achievements in Similar Works

Speech recognition topics are widely discussed in literature and scientific papers. In this section, we are going to present few results of other researchers, and which approaches and methods has been used, one of these researches investigates speaker-independent speech recognition with speaker-class models, as many as 500 speaker-class models are used to enable more precise modeling of speaker characteristics. Significant improvement was achieved using large-scale speaker-class modeling, the results indicate the possibility of reducing the error rate by up to 41.2% if the speaker-class model selection were successful. This means that the speaker-class model itself is very effective for speaker-independent speech recognition [27].

Other researchers went further to recognize the speech emotions using speaker-dependent classifiers; emotion recognition is done using global acoustic features of the speech. The feature extraction is based on calculation of global statistics of six speech signal parameters and their derivatives. For the evaluation of the system three classifiers were used: Bayes network, RBF network, and support vector machine (SVM). The obtained results suggest the availability of recognizing the speech emotions of a small group of people with good accuracy. Hence, further research should include evaluating the system in human–robot interaction scenarios, learn speech emotion expression of new people, and gender identification [28].

Berlin Emotion Database was used in another research, having the training subset as 90%, and 10% as the test subset, the feature combination of energy and pitch has the worst recognition rate, which can only recognize one emotional state. The accuracy rate for the feature combination of MFCC and mel energy dynamic coefficients (MEDC) is higher compared with the first mode. It can better recognize three standard emotional states. Adding the linear prediction cepstral coefficients (LPCC) feature, caused the performance of the model to become lower. The best feature combination is MFCC + MEDC + energy, for which the cross-validation rate can be as high as 95% for nonreal-time recognition [29].

Speaker identification took its important place in research, with different approaches to gain the appropriate goal. One of these researches was undertaken in Finland to study the role of the vector quantization in the speaker identification system, comparing the performance of different clustering algorithms, and the influence of the codebook size to find out which method provides the best clustering result, and whether the difference in quality contribute to improvement in recognition accuracy of the system.

The speaker identification using small dataset # Speakers 25 (14 M + 11 F), with average duration of training utterance 66.5 s, and average duration of testing utterance 17.7 s, with some combination the research reached 100% accuracy ratio [30].

One of the different ways of thinking came with modeling the speaker identity based on the nonlinear properties of the speech samples, where the speaker identification experiments are conducted based on the phase space distribution patterns. These are derived from the reconstructed phase space (RPS) of the speech signal named phase space point distribution (PSPD). The PSPD features obtained from five vowels are used for speaker identification purpose using the feed-forward multi-layer perceptron (FFMLP). The experiment is repeated by taking different combination of PSPD, MFCC, pitch, and first formant frequency. The experimental results indicate that the proposed phase space approach by itself is still below (31.60%) that of MFCC features (46.21%). The results further show that the combined approach in which the PSPD features, when used with MFCC, pitch, and first formant frequency, offers enormous improvement in speaker identification (on an average of 83.40%) accuracy [31].

Another case of research to improve speaker identification in environments with interfering speakers is usually the case in forensic ambient recordings. Using only the state-of-the-art MFCC–Gaussian mixture model (GMM) approach, only one of the speakers can be identified properly; on the other hand, when the independent component analysis (ICA) algorithm is applied, the identity of the $N$ speakers present on the recording, estimated with a considerable success rate. The proposed ICA-based approach requires a 10 dB less noisy audio input, but gives satisfactory results identifying all the speakers present in the mixture audio.

The database contains seven speakers, with around 2 or 3 min of speech to use as the training speech signal, and 25–40 s to use as testing speech signal [32].

Methods to identify speakers sometimes exceed the biometric models, one of the out of the box thinking depends on the pronounced names, as a source of names, plus the names written in a title block in the image track in TV broadcasts, which identify co-occurring speech turns with a very high precision.

These methods tested on the REPERE corpus phase 1, containing 3 h of annotated videos, where this system reaches an $F$-measure of 73.1%, by comparison, a monomodal, supervised speaker identification system with 535 speaker models trained on matching development data and additional TV and radio data only provided a 57.2% $F$-measure [33].

Speaker adaptation is commonly used to compensate speaker variation in large vocabulary continuous speech recognition. In a multispeaker environment where speakers change frequently, speaker segregation is needed to divide the input audio stream to speaker turns.

Speaker turns define the current speaker at each time and speaker adaptation can thus be done based on speaker turns. Speaker segregation and speaker adaptation significantly improve system performance in speech recognition task. Results of speech recognition based on testing with Finnish TV news audio, suggest that speaker adaptation may actually benefit from automatic segregation, where the speaker adaptation reduced the average letter error rate by 25% relative to baseline, automatic segregation improved correctly partitioning and labeling the speech data [34].

Another research based on deep NN (DNN) focuses mainly on locating speaker change points, where audio or speaker change point detection is the process of locating time points (or frames) in an audio stream that correspond to a transition from one speaker to another, or from music to speech or vice versa. These change points have many uses including speaker diarization, scene analysis systems or tracking speakers in a conversation, and so on.

In this research, the DNNs for change point detection are trained with 16.8 h of acoustic data, and shows that substituting the change points from KL2 metric by the change points from DNNs for speaker diarization results in a lower DER for both an English and a French diarization system [35].

For speaker diarization purposes, other systems suggests taking into consideration the visual aspects and not only the audio voices, but also visual aspects such as face and chest tracking, and lips activity detection, causing the identification to became more and more accurate.

One of these systems proposed a multimodal speaker diarization system exploiting kernel methods. Diarization is the process of partitioning an input audio stream into homogeneous segments according to the speaker identity. The focus has been put on talk-show programs, as such TV content raises challenging research issues. Two schemes are considered: an audio-only classification scheme and a parallel audio/visual classification scheme. In the latter, the classifier output is chosen through an audiovisual coherency analysis which checks if the person talking appears on-screen. If this is the case, the speaker label output by the visual classifier is chosen, otherwise the audio one is kept. Results show that, the exploitation of visual cues (when available) has been shown to be very valuable for the task of speaker diarization, even though the audio modality is the only one which is always reliable. This is owing to the fact that the active speaker is not always seen on-screen [36].

Another research proposed to perform audiovisual speaker diarization within short scenes of TV series, visually hypothesized as dialogues involving two characters. Speaker diarization is first performed separately for audio and visual features of the utterances by using the p–median model, before both resulting bipartitions of the utterance set are optimally matched in new clusters corresponding to cases of agreement between both modalities. The remaining utterances for which both modalities disagree are then acoustically assigned to the closest medoid of the newly created clusters, expected to be more robust than those based on an audio-only approach. The experimental results obtained by using both modalities turn out to outperform those obtained by purely mono-modal approaches [37].

## 5.6  Achievements in Voice Recognition

We will introduce our achievements in voice recognition. These results are obtained by applying TM and its modifications, as well as by using other methods. Presenting these results will be ordered according to the task complexity, starting with the simple case of recognition uttered words, next with the identification both speakers and words, then with the suggested text-independent speaker identification system, and finally with the speaker identification and verification systems.

### 5.6.1  The Simplest Case, Uttered Words Recognition

Many researchers started testing their algorithms and theoretical approaches with simple task, in voice recognition such a task could be recognizing very small vocabulary, usually separated words or letters.

Here, we will introduce one of these simple cases, in which the authors dealing with 11 uttered words, to identify and recognize them, however the complexity was in the database size, where the authors enlarged it to make this simple task more realistic. In any case, the authors working on visualizing the voice signal, we mean illustrating the voice signal graphically, and testing the application of Toeplitz matrices and their minimal eigenvalues together with a number of different types of NNs. The efficiency of the used methods still is very important to solve complexity of the systems, and to go further in such a research.

*5.6.1.1 Input Samples and Preprocessing Stage*   Our base consists of recorded voices for 20 people from 6 different countries, not only Arabic, adults between 23 and 45 men and women with 2 male and female teenagers. Most of the speakers are native Arabic speaking, but some are Arabic speaking European people. The total number of the recorded samples is 5472 divided into two groups. For each person and word we choose 5 samples to be the test set (totally 1100 samples), while the remaining samples (4372 samples) are chosen to be the teaching set, the ratio of training testing samples is about 80/20.

Each sample is recorded and saved in separate files; each recorded file contains only one voice with a silence region before and after the right signal. The authors used the standard format PCM, with frequency 22,050 Hz, 16-bits mono, as a standard in this work. Figure 1.3 shows a simple graph of speech preparation for feature extraction and classification.

More information and details about how to prepare the signals to recognition can be found in [8–10].

*5.6.1.2 Experiments and Result* Power spectrum estimation of Burg's model presented good results for the purpose of presenting waveform graphically; in our tests, it was used alone or as input method for other algorithms, however its parameters always had a deciding effect on the results. In our experiments, we tested all possible combinations for the following values of Burg's parameters:

The length of FFT (NFFT): 32, 64, 128, 256, 512, 1024.
The prediction order (*P*): 8, 10, 12, 16, 20, 24, 28, 32, 40.

That means we had about 54 individual situations for each method, but we will present only the best results for every method.

Table 5.2 shows the best recognition rate for each of the mentioned algorithms and the adjacent method of classification. We are not presenting the detailed parameters of the obtained results as they were given in detail in previous authors' works [8].

Table 5.2 shows how the minimal eigenvalues algorithm has not given more than 86.45% in best cases when used in its original version

**Table 5.2** Recognition Results with Different Methods of Processing and Classification

| PARAMETERS | RECOGNITION RESULTS, CLASSIFICATION BY | | | | | |
|---|---|---|---|---|---|---|
| | CONVENTIONAL METHOD | | RADIAL NN | | PROBABILISTIC NN | |
| METHOD | RECOGNIZED SAMPLES IN 1100 | SUCCESS RATE (%) | RECOGNIZED SAMPLES IN 1100 | SUCCESS RATE (%) | RECOGNIZED SAMPLES IN 1100 | SUCCESS RATE (%) |
| Burg's | 70 | 93.64 | 28 | 97.45 | 19 | 98.72 |
| Eigenvalues algorithm | 149 | 86.45 | 46 | 95.82 | 673 | 38.82 |

and without NNs. However, when used with radial NNs [11] the rate of recognition was improved to 95.82% which is very close to the efficiency of Burg method, the approach which has furnished the highest recognition rate (98.72%) when applied with the PNNs. The latter in turn proved to be impractical to use with the minimal eigenvalues or any of its modifications as it gave a much lower recognition rate than the classical classification methods, in many other cases radial NN working with eigenvalues algorithm or its modifications increased the efficiency by 30% more than the classical methods of classification.

### 5.6.2 Voiceprint and Security Systems

This section addresses the voice as an important part of biometrics, showing how it could be used with security systems. Experiments show high performance recognition rate for both uttered words and speaker. Security system build on three digits password can be easily implemented with high performance rate.

Voiceprint has many useful tasks, and can be used for different kinds of speaker and speech systems (SAS) [38–40].

This work was tested using the same database and presetting described in the previous section. The performance of the recognition system shows how the right voice and its correct speaker are identified from a spoken three-digit password. The developed model introduces a simple-to-use security system—it has two kinds of protection: the spoken digits and their speaker. The system identifies first the speaker and then the spoken password using only three spoken digits.

Studying and evaluating the effectiveness of our algorithms by demonstrating only one spoken digit, presents four possibilities for each spoken word:

1. *Success rate*: Correct recognition of the spoken word and/or the speaker.
2. *Miss word rate or miss speaker rate*: Wrong recognition of word or speaker, respectively.
3. *False speech rejection*: Correct speaker voice identification but wrong spoken words recognition.
4. *Miss speaker identification*: False rejection or false acceptance of the speaker but still recognizing the right uttered word.

**Table 5.3** Different Possibilities for Rates of Recognition System

|  | SPEAKER | |
| --- | --- | --- |
| WORD | TRUE | FALSE |
| True | Success rate | Miss speaker identification |
| False | False speech rejection | Miss word rate or miss speaker rate |

Table 5.3 explains these possibilities.

Having these possibilities, we can now proceed to perform two kinds of human–computer communication:

> *Part a—speaker identification*: Here we use the spoken words only to identify the speaker, without necessarily trying to recognize the spoken words. The efficiency of the algorithm will be higher in spite of decreasing the security level. This is very important especially when we know the algorithm is classifying the speaker from a concrete uttered word but at the same time without trying to verify if the word is the right one or not.
>
> *Part b—multilevel security for both the spoken words and their speaker*: Here we use the spoken word as a password after identifying the speaker. This means identifying the speaker first and then using his spoken words as a password. The identification rate will be lower while the level of security will be higher.

*5.6.2.1 Performance of the Speaker Identification Security System*   Table 5.4 shows the results of this experiment, where we are experiencing the right speaker identification regardless of his speech. Through different techniques of classification with Burg's method in speaker identification, the number of wrongly recognized samples was reduced to 13 in 1100 samples to have a miss rate of only 1.18%. The achieved successful recognition rate 98.82% by this method was with the PNN classifiers as can be observed in Table 5.4, which shows all different conditions. Notice that the input data to the NN are applied directly from Burg's graphs.

The obtained results showed a success rate of 94.82% by means of classical classification methods, which are based on point-to-point matching to distinguish between the examined signal features and the features of the signal-images taken from the teaching set in the database, and then classifying the signal to the most similar class.

**Table 5.4** Performance of the Speaker Identification System

| PARAMETERS | RECOGNITION RESULTS, CLASSIFICATION BY | | | | | |
|---|---|---|---|---|---|---|
| | CONVENTIONAL METHOD | | RADIAL NN | | PROBABILISTIC NN | |
| METHOD | RECOGNIZED SAMPLES IN 1100 | SUCCESS RATE (%) | RECOGNIZED SAMPLES IN 1100 | SUCCESS RATE (%) | RECOGNIZED SAMPLES IN 1100 | SUCCESS RATE (%) |
| Burg | 1043 | 94.82 | 1069 | 97.18 | 1087 | 98.82 |
| | NFFT 256, $P$ 20 | | NFFT 64, $P$ 28 | | NFFT 512, $P$ 28 | |
| TM | 956 | 86.91 | 1075 | 97.73 | | |
| | NFFT 128, $P$ 28 | | NFFT 64, $P$ 20 | | | |

The use of the NN has increased the success rate to 97.18% in the case of radial NNs and to 98.82% by the probabilistic neural ones.

Identifying the right speaker was varying from 87.27% to 100%, the average identification ratio for the 20 speakers in our database was 94.82%.

*5.6.2.2 Multilevel Security for the Spoken Words and Speaker*  Here, the experiments will deal with the second kind of human–computer communication in our suggested system. Here, we are using the spoken word as a password after having identified the speaker.

Table 5.5 shows the results of this experiment. For NFFT = 256 and $P = 20$ for Burg's method and through the classical method of

**Table 5.5** Performance Multilevel Security for the Spoken Words and Speaker

| PARAMETERS | RECOGNITION RESULTS, CLASSIFICATION BY | | | | | |
|---|---|---|---|---|---|---|
| | CONVENTIONAL METHOD | | RADIAL NN | | PROBABILISTIC NN | |
| METHOD | RECOGNIZED SAMPLES IN 1100 | SUCCESS RATE (%) | RECOGNIZED SAMPLES IN 1100 | SUCCESS RATE (%) | RECOGNIZED SAMPLES IN 1100 | SUCCESS RATE (%) |
| Burg | 1012 | 92 | 994 | 90.36 | 1072 | 97.45 |
| | NFFT 256, $P$ 20 | | NFFT 128, $P$ 28 | | NFFT 512, $P$ 20 | |
| TM | 898 | 81.64 | 1041 | 94.64 | | |
| | NFFT 128, $P$ 28 | | NFFT 64, $P$ 20 | | | |

classification, the number of correctly recognized samples for both spoken words and their speaker was 1012 in 1100 samples (92%). This speech-and-speaker recognition average rate achieved when uttering a word results in recognizing both the correct word and its correct speaker successfully, as can be seen in Table 5.5.

However, adding the NNs as classifying tools for TM–NN system has increased the success rate to an absolutely higher value (speech recognition to 95.64% and speaker recognition to 97.73%).

Not only have the Toeplitz matrix minimal eigenvalues rapidly improved the system performance, but they also have led to a very high success rate (94.64%) when used with the RBF NNs in the hybrid manner [18].

### 5.6.3 Text-Independent Speaker Identification

The voice contains many useful parts of information about the speaker, including age, gender, emotions or level of stress, and tiredness. The most important idea is the possibility of identification of a particular person using his voice samples. The voice characteristics of people are unique enough to make an attempt of identification; such methods are suitable in situations where the person has to be identified using random voice samples, for example, eavesdropping on criminals.

Speaker identification systems deals usually with real cases, sometimes it is part of security system, in other cases it is prepared to be part of more complicated systems. Here, we present algorithms for speaker identification, the proposed method considers the situation on continue speech, with fast and shortened signal processing steps.

The proposed methods demonstrate the possibility of speaker identification on the basis of continuous speech. Such ideas can be a tool for radio and TV software for recording purposes. Advanced applications can be used for monitoring and security systems based on telephony technologies as well.

#### 5.6.3.1 Database and Preprocessing
To test the algorithm in a real case we use recorded videos from Al Jazeera channel broadcast, the recorded voices belongs to many speakers, and often one speech belongs to more than one real TV session or program. Moreover, fragments with no voice data were eliminated, as mentioned above.

For every person we take a part for training purposes and the rest to be used for testing, then came the preprocessing stage. This stage aims to be simple for practical implementation, first we resample the input signal into 10 KHz (no noise reduction, no filtering, etc.). Next, we take a large segment every 2 s; each segment divided into small blocks (25 ms). Then we calculate the average energy for these blocks together and take as a real signal only these blocks with energy larger than the average (Figure 5.5).

*5.6.3.2 First Attempt*  As mentioned above, the speech samples were recorded from the known Al Jazeera news broadcasting. In order to obtain a regular database we first worked on the recorded video. The voice was then extracted, collected, and separated for each speaker. The high number of recoded video hours was reduced to about 3 h of concentrated samples of 15 different speakers, the final database consists of about 3 h of concentrated samples of 15 different speakers. Moreover, fragments with no voice data were eliminated. For every person the first 2 min were taken for training purposes and the rest was used for testing, the ratio of training–testing samples is about 17/83. A description of a selected group of samples is presented in Table 5.6.

In this case of text-independent speaker identification, the training samples were used to train the NN, later on the testing samples were

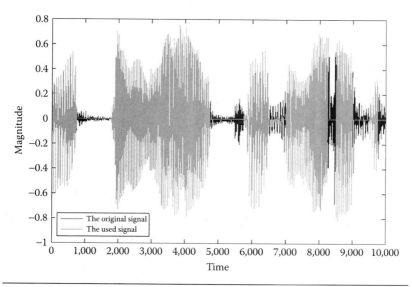

**Figure 5.5**  Obtained signal after eliminating low energy points.

**Table 5.6**   Selected Group of Voice Samples

| DATABASE OF 15 SPEAKERS | DURATION | TRAINING | TESTING |
|---|---|---|---|
| Total | 2 h:53 min:31 s | 0 h:30 min:00 s | 2 h:23 min:31 s |
| Time percentage (%) | 100 | 17.29 | 82.71 |

**Table 5.7**   Recognition Rate for Speaker Identification Using Different Methods

| | RECOGNITION RESULTS | |
|---|---|---|
| MODEL | WITH RADIAL NEURAL NETWORKS (%) | WITH PROBABILISTIC NEURAL NETWORKS (%) |
| Burg | 76.3 | 72.7 |
| Toeplitz model | 65.1 | — |

passed through the NN to testify the efficiency of the proposed algorithms, and the final results presented in Table 5.7.

*5.6.3.3 Another Attempt*   In this attempt, we used expanded voices database, recorded from Al Jazeera Arabic news broadcasting. The recorded samples were for 30 different people with recorded periods from 3 min:4 s to 19 min:31 s, and often in different sessions for the same speaker; the sum of the recorded samples were exactly 3 h:57 min:33 s, the recorded samples in some cases were transmitted by phone but often recorded in professional studios.

We divided the database into training and testing sets. For each speaker we take 2 min to the training set, 30 min for the whole database, and the rest of data for the testing set, the ratio of training testing samples is about 13/87.

The result presents the average of the right identification of the speakers. The standard format used by the authors is PCM, with a frequency of 22,050 Hz, 16-bits mono. Each file presents an individual speaker, and for many speakers the samples were recorded in different sessions (the reason that caused the algorithm to fall down in many cases).

In this research we conducted a wide variety of experiments using all the mentioned approaches in this chapter. However, here and for simplicity we will present only those with the best results leaving the others for future presentation when improving and updating the work on them. Table 5.8 shows the best recognition rate for each of the listed algorithms.

**Table 5.8**   Results of Speaker Identification Recognition Rate According to the Attempt in Section 5.6.3.3

| | RECOGNITION RESULTS | |
|---|---|---|
| MODEL | WITH RADIAL NEURAL NETWORKS (%) | WITH PROBABILISTIC NEURAL NETWORKS (%) |
| Burg | – | 77.4 |
| Toeplitz model | 72.6 | – |
| MFCC | – | 84.6 |

*5.6.4  What about Speaker Verification?*

Speaker verification is an important part of any security system. To verify the possibility of speaker verification, the authors used the SVMs. Meanwhile the base algorithms for feature extraction are still Burg, Toeplitz algorithm, and MFCC, for classification NNs, and the new step is to try verifying the results using SVMs.

These tests have been made using another voice database recorded manually, for testing purposes. The prepared database was recorded from Al Jazeera news broadcasting, in PCM format, with frequency 48 kHz, 32-bits, stereo.

This database has been called base database, consisting of three females, and nine males, the extra database has been used for verification purposes based on four male voices. Using the base database: total 12 Speakers 1 h:22 min:43 s, the following time periods were obtained:

*Training set*: 0 h:20 min:41 s (25% of 1 h:22 min:43 s)
*Testing set*: 1 h:02 min:02 s (75% of 1 h:22 min:43 s)
*Extra database*: Four speakers
*Testing set*: 0 h:20 min:40 s
Each sample consists of 3 s signal

Table 5.9 shows the details of the used database.

**Table 5.9**   Structure of Voice Database

| Database (16 speakers) | | |
|---|---|---|
| 1 h:43 min:23 s | | |
| Base (12 speakers) | | Extra (4 additional speakers) |
| 1 h:22 min:43 s | | 20 min:40 s |
| Training set | Testing set | |
| 20 min:41 s | 1 h:02 min:02 s | |
| 25% of (1 h:22 min:43 s) | 75% of (1 h:43 min:23 s) | |

*5.6.4.1 Identification Treatment* In this experiment, we are trying to answer standard classification question in pattern recognition, which is: what is the best model for this pattern? (See Figure 5.6.)

Build NN, and use it to classify all the testing sets, as used in many earlier researches.

Such kind of work has been widely tested by authors in different publications, and here we will use the base database for training and testing, and the results are promising (Table 5.10).

*5.6.4.2 Verify the Speaker—Claiming It Correctly* Here we try to answer the question: Does this pattern belong to this speaker model? To get the results we needed to build a SVM model for each speaker; by

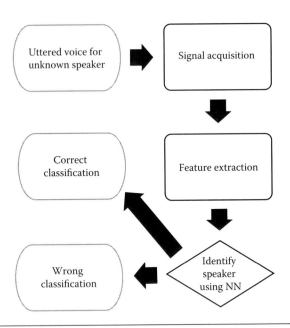

**Figure 5.6** Workflow diagram for the identification process.

**Table 5.10** Results Obtained Solving Identification Problem

| | RECOGNITION RESULTS | |
| --- | --- | --- |
| MODEL | WITH RADIAL NEURAL NETWORKS (%) | WITH PROBABILISTIC NEURAL NETWORKS (%) |
| Burg | – | 88 |
| Toeplitz model | 78 | – |
| MFCC | – | 85 |

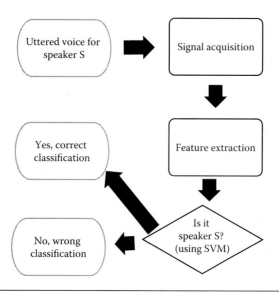

**Figure 5.7** Workflow diagram of the verification experiment, showing that the speaker is recognized correctly?

**Table 5.11**   Results Solving Verification Problem for the Case in Section 5.6.4.2

| MODEL | VERIFICATION RESULTS (%) |
| --- | --- |
| Burg | 82 |
| Toeplitz model | 77 |
| MFCC | 93 |

the results the vector should be able to answer the former mentioned question (Figure 5.7).

In this case, the results present the true acceptance ratio, which means that the verification process succeeded in verifying the right speakers, and the results are tabulated in Table 5.11.

Of course, a real system will consist of the two previous tests, having identification and verification as well, with combined results; rates then will go down but still in a good range.

*5.6.4.3 True Rejection and False Acceptance*   Having to answer the question: Does this pattern belong to this speaker model? We try to find the false acceptance, and the true rejection ratios—using the database for training samples, claiming the speaker wrongly (Figure 5.8).

This kind of verification is also very important when we seek for high standard of security systems, and the results are tabulated in Table 5.12.

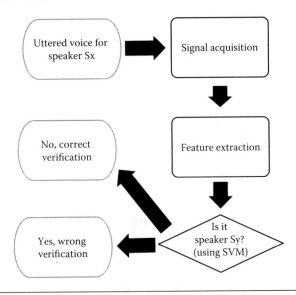

**Figure 5.8** Workflow diagram of the true rejection and false acceptance issue.

**Table 5.12** Results Solving Verification Problem for True Rejection and False Acceptance Case in Section 5.6.4.3

| MODEL | VERIFICATION RESULTS (%) |
|---|---|
| Burg | 2.5 |
| Toeplitz model | 2.6 |
| MFCC | 0.75 |

False acceptance ratio and true rejection ones are in good state, with significant advantage of MFCC algorithm.

*5.6.4.4 Extra Testing Data for Verification*　In this verification issue, we try to answer the question: Does this pattern belong to this speaker model? The last try will be for external data, for patterns that do not exist in training sets, and here we can just verify the false acceptance ratio, but here with data belongs to speakers outside the training set.

The training data will be base database training data, for which we build SVM model for each speaker, and the testing will be the extra database, not used in training SVMs. The verification process is presented in Figure 5.9.

This case was different from any older experiments the authors made. The given results gave a high ratio of false acceptance for speakers not included in the training sets. Table 5.13 shows the results obtained in this case.

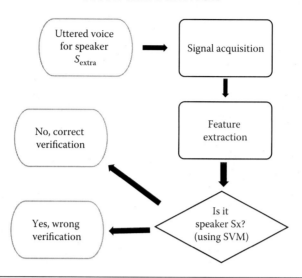

**Figure 5.9**   Workflow diagram of the false acceptance issue, using speech recorded for speakers outside the base database.

**Table 5.13**   Results Obtained Testing the True Acceptance Verification Problem, Using Speakers from the Extra Database

| MODEL | VERIFICATION RESULTS (%) |
|-------|--------------------------|
| Burg | 11 |
| Toeplitz | 10 |
| MFCC | 10 |

The last results are unsatisfied, taking into consideration the verification issue, the SVM model has been expected to give clear decision about throwing the untrained patters, and to classify it out of the other data; however, we need to review the algorithm and to test other aspects related to it.

As a summary, different methods have been presented for voice recognition cases. Many of them are very promising. Meanwhile, the authors are still working on developing the mentioned approaches and algorithms to obtain much better results.

## 5.7 Conclusions

Voice recognition feature is one of the most attractive behavioral biometrics example according to many research teams and institutions as well as industrial sides. As the only behavioral feature that has physiological aspects, it can always be counted on for its higher possible recognition rate. In this meaning, the repeatability is higher leading to high efficiency results.

This chapter was devoted to speech and speaker recognition using mainly the authors' approaches. The development of the worked out methods and algorithms is presented together with the practical results gained for years. As the results have shown, there were even trials for user identification (not only verification) by voice recognition. Although the results were promising, we still are seeking other methods to use voice as one of 2–3 models in multimodal systems.

# References

1. J. P. Campbell, Speaker recognition: A tutorial. *Proceedings of the IEEE*, 85(9), 1437–1462, 1997.
2. J. R. Vacca, *Biometric Technologies and Verification Systems*, 1st edn., Butterworth-Heinemann, Amsterdam, the Netherlands, 2007.
3. M. Al-Akaidi, *Fractal Speech Processing*, Cambridge University Press, Cambridge, 2004.
4. M. Alghamdi, *Arabic Phonetics*, Attaoobah, Library, Saudi Arabia (in Arabic), 2000.
5. J. Anthony Seikel, D. W. King, and D. G. Drumright, *Anatomy & Physiology for Speech, Language, and Hearing*, 4th edn., Delmar Cengage Learning, Clifton Park, NY, 2009.
6. K. Saeed, *Image Analysis for Object Recognition*, Bialystok Technical University, Poland, 2004.
7. K. Saeed and M. Kozłowski, An image-based system for spoken-letter recognition. In *10th International Conference on Computer Analysis of Images and Patterns*, August 2003, Groningen, the Netherlands. Lecture Notes in Computer Science, vol. 2756, Springer-Verlag, Heidelberg/Berlin, Germany, pp. 494–502, 2003.
8. K. Saeed and M. K. Nammous, Heuristic method of Arabic speech recognition. In *7th International Conference on Digital Signal Processing and Its Applications*, IEEE, Moscow, Russia, pp. 528–530, 2005.
9. K. Saeed and M. Tabędzki, Intelligent feature extract system for cursive-script recognition. In *Proceedings of the 4th IEEE International Workshop on Soft Computing as Transdisciplinary Science and Technology*, Springer-Verlag, Heidelberg, Germany, Advances in Soft Computing, pp. 192–201, Muroran, Japan, 2005.
10. E. A. Guillemin, *A Summary of Modern Methods of Network Synthesis—Advances in Electronics*, vol. III, Academic Press, New York, pp. 261–303, 1951.

11. K. Saeed, Computer graphics analysis: A criterion for image feature extraction and recognition, *MGV—International Journal on Machine Graphics and Vision*, 10(2), 85–194, 2001.

12. F. H. Effertz, On the synthesis of networks containing two kinds of elements. In *Symposium on Modern Network Synthesis*, Polytechnic Institute of Brooklyn, New York, pp. 145–173, 1955.

13. K. Saeed, Carathéodory–Toeplitz based mathematical methods and their algorithmic applications in biometric image processing. *Applied Numerical Mathematics*, 75, 2–21, 2014.

14. R. Tadeusiewicz, *Speech Signals*, WKiL (in Polish), Warsaw, 1988.

15. J. Durbin, Efficient estimation of parameters in moving average models. *Biometrics*, 46(3–4), 306–316, 1969.

16. V. K. Ingle and J. G. Proakis, *Digital Signal Processing Using MATLAB*, Brooks Cole, Pacific Grove, CA, July 1999.

17. D. Rocchesso and F. Fontana (eds), *The Sounding Object*, Mondo Estremo Publishing, Italy, 2003.

18. K. Saeed and M. Nammous, A speech-and-speaker identification system: Feature extraction, description and classification of speech-signal image. *IEEE Transactions on Industrial Electronics*, 54(2), 887–897, April 2007.

19. K. Saeed and M. Nammous, A new step in Arabic speech identification: Spoken digit recognition. In *Information Processing and Security Systems*, K. Saeed and J. Pejaœ (eds), Springer Science+Business Media, New York, pp. 55–66, 2005.

20. S. B. Davis and P. Mermelstein, Comparison of parametric representations for monosyllabic word recognition in continuously spoken sentences. *IEEE Transactions on Acoustics, Speech, and Signal Processing*, 28, 357–366, 1980.

21. L. C. W. Pols, Spectral analysis and identification of Dutch vowels in monosyllabic words, PhD thesis, Free University, Amsterdam, the Netherlands, 1977.

22. M. D. Skowronski and J. G. Harris, Exploiting independent filter bandwidth of human factor cepstral coefficients in automatic speech recognition. *The Journal of the Acoustical Society of America*, 116(3), 1774–1780, Sept. 2004 [Online]. Available: http://www.cnel.ufl.edu/~markskow/papers/hfccJasa.pdf (accessed on July 2, 2016).

23. T. Ganchev, N. Fakotakis, and G. Kokkinakis, Comparative evaluation of various MFCC implementations on the speaker verification task. In *10th International Conference on Speech and Computer*, Patras, Greece. Vol. 1, pp. 191–194, October 17–19, 2005. [Online]. Available: http://citeseerx.ist.psu.edu/viewdoc/download?doi=10.1.1.75.8303&rep=rep1&type=pdf (accessed on July 2, 2016).

24. M. M. El Choubassi, H. E. El Khoury, C. E. Jabra Alagha, J. A. Skaf, and M. A. Al-Alaoui, Arabic speech recognition using recurrent neural networks. In *Proceedings of IEEE International Symposium on Signal Processing and Information Technology*, IEEE, Darmstadt, Germany, 2003.

25. T. Hill and P. Lewicki, *Statistics Methods and Applications*, StatSoft Inc., Tulsa, OK, 2006.

26. M. W. Mak, W. G. Allen, and G. G. Sexton, Speaker identification using radial basis functions. In *3rd International Conference on Artificial Neural Networks*, University of Northumbria, Newcastle, UK, 1998.

27. K. Konno, M. Kato, and T. Kosaka, Speech recognition with large-scale speaker-class-based acoustic modeling. In *Signal and Information Processing Association Annual Summit and Conference, Asia-Pacific*, IEEE, Kaohsiung, China, pp. 1–4, Oct. 29–Nov. 1, 2013.

28. L. Juszkiewicz, Improving speech emotion recognition system for a social robot with speaker recognition. In *19th International Conference On Methods and Models in Automation and Robotics*, IEEE, Miedzyzdroje, Poland, pp. 921–925, Sept. 2–5, 2014.

29. P. Shen, Z. Changjun, and X. Chen, Automatic speech emotion recognition using support vector machine. In *International Conference on Electronic and Mechanical Engineering and Information Technology*, IEEE, Harbin, Heilongjiang, China, pp. 621–625, Aug. 12–14, 2011.

30. T. Kinnunen, T. Kilpelainen, and P. Franti, *Comparison of Clustering Algorithms in Speaker Identification*, University of Joensuu, Finland, 2000.

31. V. L. Lajish, R. K. Sunil Kumar, and P. Vivek, Speaker identification using a nonlinear speech model and ANN. *International Journal of Advanced Information Technology*, 2(5), 15–24, October 2012.

32. M. A. Silveira, C. P. Schroeder, J. P. C. L. da Costa, C. G. de Oliveira, J. A. Apolinário, Jr, A. M. Rubio Serrano, P. Quintiliano, and R. T. de Sousa, Jr, Convolutive ICA-based forensic speaker identification using Mel frequency cepstral coefficients and Gaussian mixture models, *The International Journal of Forensic Computer Science—IJoFCS*, 8(1), 27, 2013.

33. J. Poignant, L. Besacier, and G. Quénot, Unsupervised speaker identification in TV broadcast based on written names. *IEEE/ACM Transactions on Audio, Speech, and Language Processing*, 23(1), 57–68, Jan. 2015.

34. U. Remes, J. Pylkkonen, and M. Kurimo, Segregation of speakers for speaker adaptation in TV news audio. In *IEEE International Conference on Acoustics, Speech and Signal Processing, ICASSP*, IEEE, Honolulu, HI, pp. IV-481–IV-484, April 15–20, 2007.

35. V. Gupta, Speaker change point detection using deep neural nets. In *IEEE International Conference on Acoustics, Speech and Signal Processing*, IEEE, South Brisbane, Australia, pp. 4420–4424, April 19–24, 2015.

36. F. Vallet, S. Essid, and J. Carrive, A multimodal approach to speaker diarization on TV talk-shows. *IEEE Transactions on Multimedia*, 15(3), 509–520, April 2013.

37. X. Bost, G. Linares, and S. Gueye, Audiovisual speaker diarization of TV series. *IEEE International Conference on Acoustics, Speech and Signal Processing*, IEEE, South Brisbane, Australia, pp. 4799–4803, April 19–24, 2015.

38. M. Alghamdi, Speaker's identification—Voice print. *King Khalid Military Academy Quarterly*, 14(54), 24–28, 1997 (Saudi Arabia [in Arabic]).

39. J. Lindh, Handling the voiceprint issue. In *Proceedings of FONETIK 2004*, Department of Linguistics, Göteborg University, Sweden. [Online]. Available: http://www.ling.su.se/fon/fonetik_2004/lindh_voiceprint_fonteik2004.pdf (accessed on July 2, 2016).

40. G. B. Mindlin, M. Trevisan, and M. Eguia, Topological voiceprints for speaker identification, UCSD-UBA-UNQ Patent application filled by UCSD, application number 60/497,007 Priority date, Aug. 20, 2003. Aug. 20, 2004. [Online]. Available: http://www.df.uba.ar/~marcos/voicePHYSD.pdf (accessed on July 2, 2016).

# Index

Note: Page numbers followed by f and t refer to figures and tables, respectively.

identification treatment, 208,
208f, 208t
true rejection and FA,
209–210, 210f, 210t
*vs.* speech recognition, 17
Speaker-dependent voice
recognition, 16–17
Speaker-independent voice
recognition, 16–17
Spectral analysis, 188
Speech recognition, 20, 184
achievements, 195–199
*vs.* speaker recognition, 17
Stance phase, 142–143
State-of-the-art method, 143
Subgait, 152
Support vector machine (SVM),
77–78, 146, 195, 207,
210–211
SVC2004. *See* Signature Verification
Competition (SVC2004)
SVM. *See* Support vector machine
(SVM)
Swing phase, 142

**T**

Technical University of Munich,
Germany, and Indian
Institute of Technology
Kharagpur (TUM-IIT
KGP) gait dataset, 156
Telegraph, 93–94
Tensor decomposition, 151
Testing phase, partial silhouette, 168
Text-dependent speaker
identification system, 16
Text-independent speaker
identification, 16, 204–207
another attempt, 206, 207t
database and preprocessing,
204–205, 205f
first attempt, 205–206, 206t

Timing data, keystroke features, 96
Toeplitz matrix minimal eigenvalues
algorithm, 189–193
advantage, 190
linear predictive coding and
Burg's model, 190–192, 192f
MFCC, 192–193
Traced forgery, 47
Training phase, partial silhouette,
166–168
Trajectory reconstruction, 61f, 62
True positive (TP), 27
True rejection, 209–210, 210f, 210t
TUM-IIT KGP (Technical
University of Munich,
Germany, and Indian
Institute of Technology,
Kharagpur) gait dataset, 156
TypeWATCH, 132
Typing features, 111–113
Typing speed, keystroke, 96

**U**

UMD (University of Maryland) gait
dataset, 156
Uniqueness, keystroke dynamics, 106
Universality, keystroke
dynamics, 106
Universal object recognition system,
21, 21f
University of Maryland (UMD) gait
dataset, 156
University of South Florida (USF)
gait dataset, 155–156
User authentication methods,
103–104
User authorization, keystroke
dynamics, 101
USF (University of South Florida)
gait dataset, 155–156
Uttered words recognition,
199–201, 200t